THE LATE RICHARD BENNETT.

HISTORY OF CORN MILLING

BY

RICHARD BENNETT AND JOHN ELTON

VOL. IV

SOME FEUDAL MILLS

WITH MEMOIR OF THE LATE RICHARD BENNETT

By JOHN ELTON

Dee Mills, Chester.

EP Publishing Limited
1975

Republished 1975 by EP Publishing Limited
East Ardsley, Wakefield
West Yorkshire, England

First published 1904 by Simpkin, Marshall and Co., Ltd

ISBN 0 7158 1138 x

Please address all enquiries to
EP Publishing Limited (address as above)

Printed in Great Britain by
Redwood Burn Limited
Trowbridge & Esher

PREFACE.

THE short histories of mills comprising this concluding volume are intended to illustrate the actual operation of the various laws and customs already recited in detail. They recount how, in accordance with those laws and customs, milling soke in each district was locally created by the superior lord; how it was maintained, by his heirs or assigns, with considerable trouble to themselves and much anxiety to the public; and how, finally, with the march of the times the force of popular hostility gradually overcame and extinguished it. This was the general fate of milling soke throughout the country, though in some few instances, as related in a previous volume, the ancient privileges had been preserved even down to the present century, and were only abolished at last by public purchase. It will be remembered that, as no general statute law had created soke, so no law at one stroke abolished it throughout the kingdom. Thus each district was thrown entirely upon its own resources to obtain relief from what, once a valuable boon, had become an intolerable yoke on the people and an impolitic hindrance to industrial and commercial progress.

Histories of corn mills constitute a somewhat novel branch of archæology. But since the issue of the first volume of this work, various local antiquaries have shown that it is an interesting and fruitful one; an excellent illustration of what may be done in this direction being afforded by a history of certain ancient mills at Glasgow, written by their owner, a practical miller, Mr. James White, F.S.A. (Scotland).*

This volume, the scheme of which was draughted in collaboration with Mr. Richard Bennett, has been written

* *The Lands and Mills of Partick.* Read before the Scottish Association, December 1898.

mainly on the lines then adopted, though, as a result of many researches undertaken later, its scope has been considerably extended; and the historical sketches now presented are believed to be as authoritative and complete as the records will allow.

Unhappily, however, this has only been accomplished since, on the eve of the publication of Volume III., the work was suddenly deprived of its patron and director: Mr. Bennett having died, almost without warning, in the prime of life, and, apparently, perfect health, from an abscess on the brain, on May 8, 1900. In now issuing the volume under circumstances rendering his participation in the work but a memory, one cannot but endeavour to preface it with some account of the earnest exertions he so long and so pleasantly put forth on its behalf; and recall a few of those events in an arduous career which, towards its untimely close, seem to have appropriately converged in the study of the present congenial theme.

It is a remarkable fact that, not to an archæologist or miller, not to a gentleman of leisured ease or great wealth, but purely to a commercial man, daily engrossed in mercantile affairs, the inception and publication of this History are due. It was Mr. Bennett, solely a man of business, who, conceiving the project, dreamed of making amends for two centuries' literary neglect of mills and milling; and who gave unstintingly labour, time, and money to the achievement of that end. To those who knew him this singular fact was no paradox. He had retained a long association with the corn trade of the port of Liverpool, and enjoyed intimate relations with its foremost millers; he possessed an inherent literary and antiquarian taste, and took unceasing interest in the fast-disappearing old mills of his native city; while over all he was animated by a strong originality of character and an indomitable spirit of enterprise. These were the factors that brought about his embarking on an archæological enterprise of this character.

He had for some years desired to collect what little gleanings of histories, reminiscences, or traditions might remain regarding the antique windmills that in his youth had abounded in Liverpool, and in the early part of 1894 he commissioned the present writer to compile a record of the kind. From so small and local a starting-point eventually originated, under his enterprising auspices, this entire History. In the earliest investigations undertaken with respect to Liverpool, the mystery surrounding soke soon fascinated his practical mind; and the lack of any work relating to the existence and incidence of so curious a monopoly struck him as an absolute hiatus in the practical knowledge of the day. Similarly, the use of early handstones and the origin and development of power mills absorbed his interest; and here again a vast dearth of information was only too evident. On all these topics, in fact, the only authorities of moment were writers whose Latin treatises, crude and imperfect, were long since obsolete.* It was thus only step by step, as investigation proceeded, that the project of this History was decided upon.

The necessary researches and literary compilation of the work being entrusted to the writer, Mr. Bennett, in the hearty, thoroughgoing manner that characterised all his doings, concerned himself with the practical side of the undertaking. No trouble, no limit of time, no expense deterred him from the task, or ever prevented him adopting any suggestion tending to extend enquiry or facilitate full study of the subject.

While investigations were in progress it became clear that, as regarded early extinct handstones, a perfect knowledge of their nature and method of working could only be secured by the actual handling and working of the stones themselves. Mr. Bennett therefore resolved to collect specimens; and possible sources whence they might be acquired being gradually ascertained, he systematically and, almost as though the affair were a matter of momentous business, followed with the utmost assiduity every clue to the desired relics. His persistent efforts in this direction were in time rewarded by the possession of a collection that in extent, variety, and representative character was, and, so far as can be traced, still is, unrivalled. About a hundred

* *De Molindis*, J. Heringius, Lyons, 1663. *De Molis Manualibus Veterum*, G. G. Hoheisel, Dantzic, 1728. *De Pistrinis Veterum*, F. Goetzius, Venice, 1730. *De Molarum Initiis: De Molarum Progressibus*, G. H. Ayrer, Gottingen, 1772.

specimens from the unique collection, showing the evolution of the various handstones and wooden grain implements, British and foreign, ancient and modern, were presented, after his death, in accordance with what had been his own desire, to the Liverpool Public Museum.* Nor did his practical co-operation in the work end here. As the compilation of the history proceeded at his residence (Stonycroft House) he regularly devoted early mornings or late evenings to collaboration—a pursuit in which he found constant recreation and rest from the business cares of the day. Six years of enthusiasm were thus devoted to the pleasant task; but the only result he lived to see was the kindly and gratifying reception accorded to the first two volumes. Since then, the third volume having been similarly honoured, the present and last of the series has been written

* "By purchase, exchange, or gift almost every desirable specimen that could be discovered in private hands was resolutely acquired. Friend nor acquaintance at home or abroad was exempt from his enquiries on this behalf, while professional collectors of curios in all parts of the world were diligently circularised. From all quarters the precious stones and utensils of wood came. From farmers of Anglesey to the governor of isles in the Pacific; from metropolitan dealers to African missionaries; from peasants of Norway to Arabs dwelling on the site of Homer's Troy; from dwellers in the Midlands of England to sojourners amid the hills of Mexico—literally 'from China to Peru'—the scattered units to form the collection were amassed; while was also added a stone certified to have been dug up by a gardener at Anne Hathaway's cottage at Stratford-on-Avon that may once have ground flour for Shakespeare."—*Milling*, May 19, 1900.

Mr. William C. Edgar, of *The North-Western Miller*, Minneapolis, has been the first authority on the subject to recognise the singular interest attaching to this difficultly acquired collection, and to desire its suitable preservation. Last autumn, having communicated with the writer as to the possibility of purchasing it, he learned of its disposal with regret. There is warrant for stating that, had his rare appreciation of its historic value been known earlier, Mr. Bennett himself would have cordially desired its acceptance by Mr. Edgar in furtherance of so great an aim as the foundation of an international Museum of Milling. The following letter exhibits Mr. Edgar's initiation of this notable project:—

"MINNEAPOLIS, *November* 2, 1903.

"DEAR SIR, . . . I very much regret to learn that the collection of millstones belonging to the late Mr. Bennett has been turned over to the Liverpool Museum. I was prepared to purchase this collection, and use it as a nucleus for the most complete museum of milling appliances ever brought together in one spot. To this end it was my intention to secure the erection in this city of a building for the purpose, and to instal therein this collection, with such milling literature, photographs, drawings, &c., as might be possible to get together, the whole making a complete milling museum. It seemed to me that Minneapolis, being the greatest milling point the world has ever known, would be the most appropriate place in which to establish an institution of this sort. I feel that I could have interested with me in this work the leading millers of the country, if not of the world; and the result would have been an institution which, in time, would have been of the utmost value to the whole milling trade. I am much disappointed to learn that the disposition made of the collection has rendered my plan quite impossible, as I would not feel equal to undertaking such a collection from the beginning.

"Very truly yours,

"WM. C. EDGAR.

"MR. JOHN ELTON,
"Birkdale, Southport."

in tribute to his memory and the hopeful aims he had in view; the publishing expenses being defrayed by his family.

In commercial circles Mr. Bennett enjoyed for years a high position, which he had attained solely by his own exertions and in face of many difficulties. The son of a working carpenter, he was born in 1844 in Christian Street, Liverpool; and at a very early age, without friends, trade, or even elementary education, sought for himself the precarious employments such a child could obtain. His foot rested on the lowest rung of the ladder, and in future years he was even content that it should have been so. It was not till, some years later, he contrived to gain a place in a warehouse near the docks that the disabilities due to a want of education dawned upon him, and thereupon he joined a night school; thus beginning a course of self-help and self-education which he steadily maintained throughout life. Having at length secured a position as clerk in the riverside forwarding business, he was ultimately engaged by Messrs. Notara, Greek corn factors and warehousemen, with whom he remained some years, and attained the position of manager. On their retirement he was enabled to commence on his own account in the same line, adding thereto the riverside carrying business; starting with an excellent reputation, if little capital, and with all his early habits of strenuousness, abstemiousness, and economy. His small business gradually developed into a large one. Years before his death the firm of Richard Bennett & Co. had risen foremost among the forwarding agents of the port, where their scores of handsome teams of heavy draught horses long constituted a notable feature of the riverside traffic; and where Mr. Bennett himself, by resourcefulness of character, business reputation, and cheerful *bonhomie*, had become one of the public men of Liverpool. Still, though several times solicited to enter the City Council, he always declined the honour.*

A man of strong domestic tastes, happily married, and with a family of young children, he had at the first

* "A good specimen of the sturdy, fearless, self-made Dicky Sam, the story of his early struggles compels the respect and admiration of even those who envy his success."—*Liverpool Porcupine*, May 9, 1896.

opportunity settled in the country, and occupied his leisure with fruit-growing and the rearing of bees. In this latter pursuit he exhibited vast enthusiasm: bees almost rivalled shire horses as a hobby; his first special collection of books consisted of old editions of works on bee culture; he became a noted bee-master, and for several years acted as secretary of the local Bee Keepers' Association. At the same time his literary bent of mind was manifested in the gradual acquirement of a good library, the special feature of which consisted of a collection of antiquarian and historical works relating to Liverpool.

In 1899 the attraction of large gardens attached to Yewdale House, Huyton, then in the market, induced him to purchase this residence; and here, during a few months, he passed perhaps the most enjoyable period of his life. Early in the spring following the insidious malady which attacked him was scarcely diagnosed, when, after but two or three days' confinement to his room, he died with great suddenness. At Anfield Cemetery, Liverpool, his interment took place in the presence of a large concourse of personal friends and representatives of public and commercial associations.*

* "His loss at the early age of fifty-six years will be felt by a very wide circle of friends as well as business acquaintances in Liverpool and elsewhere, for wherever he was known he was respected and esteemed. He was a man of great force of character, a fine type of a self-made English man of business. Only those whose privilege it was to know him intimately had any knowledge of the many fine traits in his character or suspicion of his many charitable acts. He was one of the best secretaries the Liverpool Corn Trade Guild ever elected to that post of honour, and was instrumental in raising that institution to its present high state of prosperity and usefulness. He was chairman of several business undertakings, which, greatly owing to the zeal displayed by him, prospered exceedingly; but doubtless he will best be remembered by his fine work upon the History of Corn Milling."—*Corn Trade News*, May 8, 1900.

CONTENTS OF VOL. IV.

CHAPTER I.

CASTLE MILLS, DUBLIN.

CHAPTER II.

SHREWSBURY ABBEY MILLS.

CHAPTER III.

DEE MILLS, CHESTER.

CHAPTER IV.

KING'S MILLS OF THE ROYAL BOROUGH OF LIVERPOOL.

CHAPTER V.

JEDBURGH CORPORATION MILLS.

LIST OF ILLUSTRATIONS.

SOME FEUDAL MILLS.

CHAPTER I.

THE CASTLE MILLS, DUBLIN.

I. THE CASTLE MILLS, DUBLIN.

1. Site, &c., 1248.

1. THE early Plantagenet mills beside Dublin Castle were not extensive, but some little interesting history centres round them. The following episodes appear to be all of importance that may be gathered from the available records. In connection with the somewhat earlier documents relative to the king's mills of Ardee, they illustrate the varied circumstances attendant upon the founding and working of king's mills in that period. The Castle Mills comprised but an ordinary manorial establishment, worked in accordance with civil law; not garrison mills maintained under military rule. The sovereign acted purely as manorial lord in the management of the place; and Henry III., in compounding for damages alleged by certain monks against him as founder of Castle Mills, evinced a respect for civil law and custom which feudal manorial mill-owners, generally, only too often failed to emulate.

Text, III. 4.

When Dublin Castle was founded by John in 1204, the milling of that locality was conducted by the monks of St. Mary's Abbey. This abbey, which is invariably termed St. Mary's "juxta Dublin," stood

I. THE CASTLE MILLS, DUBLIN.

1. Site, &c., 1248.

Hist. and Mun. Doc. Ireland. Chron. and Mem., liii.

outside the walls and across the Liffey, almost opposite to the Castle.

The association of the abbey with milling in the neighbourhood of the Castle is referable to an undated grant made to William Doubleday by John (before his accession to the throne) of the site of a mill there. The grant seems to have been soon after made over by William to his heir Germanus, who divided the property with the abbey: the half-share of the latter in the mill, and in the common bakery attached to it, including in its soke district the site of the new Castle. The deeds for these transactions are annexed:—

Dublin White Bk., fol. 41.

Johannes filius domini regis Anglie et domini Hibernie, archiepiscopis episcopis abbatibus comitibus baronibus justiciariis constabulariis et omnibus ballivis et ministris et fidelibus suis Francis et Anglicis et Hiberniensibus de tota Hibernia salutem.

Sciatis me dedisse et concessisse et hac presenti carta mea confirmasse Willielmo Dubelday sedem unius molendini inter portam ecclesie Sancte Marie de Dublin et ecclesiam Sancti Andree et cursum aque quantum pertinet ad predictam sedem molendini, tenendos de me et heredibus meis illi et heredibus suis per servicium unius esturci sori per annum pro omni servicio.

Quare volo et firmiter precipio quod predictus Willielmus et heredes sui post eum habeant et teneant predictam sedem molendini per predictum servicium bene et in pace libere et quiete integre et plenarie et honorifice cum omnibus libertatibus et liberis consuetudinibus que ad predictam sedem molendi pertinent.

John, son of the Lord the King of England and Lord of Ireland, to the archbishops, bishops, abbots, earls, barons, justiciaries, constables, and all bailiffs and officials, and his faithful people, French and English, and Irish of the whole of Ireland, greeting:—

Know ye that by me is dedicated and conceded and by this my present charter is confirmed to William Doubleday the site of one mill between the gate of the church of St. Mary of Dublin and the church of St. Andrew, also the watercourse so far as it appertains to the said site: he and his heirs to hold the same of me and my heirs by the presentation of one hawk per annum for all services.

Wherefore I will and firmly enjoin that the said William and his heirs after him have and hold by the said service the said mill site well and in peace, freely and quietly, wholly and fully, and with honour; with all liberties and free customs which belong to the said site.

I. THE CASTLE MILLS, DUBLIN.

1. Site, &c., 1248.

Hec est convencio facta inter abbatem et conventum Sancte Marie juxta Dublin ex una parte et Germanum filium Randulphi Dubylday videlicet quod abbas et conventus habebunt medietatem ipsius molendini et omnium ovencionum inde proveniencium: et dictus Germanus Dubylday et heredes ejus aliam medietatem et omnes ovenciones inde provenientes habebunt.

Carta vero regis quam Willielmus habuit de eodem molendino tradita est in custodia prefati Germani per quam ipsum molendinum cum opus fuerit debet defendi. Cart. Domus Beatie Marie, fol. 65.

Ut autem hec convencio firma sit et stabilis imperpetuum presens scriptum sigillis utriusque partis roboratur.

Hiis testibus Andrea Brun Turstano de Balinor Olivero de Cestria et Johanne filio ejus Willielmo Dubulday Roberto fratre ejus et multis aliis.

This is the agreement made between the abbot and convent of St. Mary juxta Dublin of the one part and Germanus, son of Randulph Doubleday: viz. that the abbot and convent shall have half of his mill and all the baking which proceeds from it [at the common oven], and the said Germanus and his heirs shall have the other half and all the baking which proceeds from that. Text, III. 207.

The original charter of the king for this mill, which William had, has been handed down to the custody of the said Germanus, by which [the title to] the said mill and its rights may be defended.

Furthermore, that this agreement may be firm and stable in perpetuity, the present deed is fortified by the seals of both parties. Witnesses—Andrew Brown, Thurstan de Balinor, Oliver of Chester and John his son, William Doubleday, Robert his brother, and many others.

This site is identified by the allusion to the churches of St. Mary and St. Andrew. It was near here that, in 1248, Henry III. established his castle mills at the foot of the eastern slope of the Castle hill, upon the small river Podder. These were within the city walls and near the gate called Gate del Dam, the closing of which during the war in 1316 gave rise to a subsequent claim by the millers. They also were near the church of St. Mary del Dam.* Apparently,

* Harris believes the name of the church to have been St. Mary la Dame; a suggestion which has probability, but lacks evidence; and disregards the existence of the contiguous mill-dam. Warburton, in 1818, adopts Harris's opinion. But this is no wonder, since in his pretentious *History of Dublin*, published in two quarto volumes, he adopts bodily, verbatim et literatim, chapter by chapter, and without acknowledgment, the whole three hundred pages of Harris's modest little tome on the antiquities of Dublin. Warburton died before this work was published, or he would doubtless have afforded some acknowledgment of a fact which his literary executors scantily admit and do not explain:—"The only history of the city hitherto published is that of Harris; its antiquities were highly valuable, and were made ample use of by Mr. Warburton."

1. Site, &c., 1248.

therefore, when this church was founded, there was already an ancient mill-dam on the stream; and St.

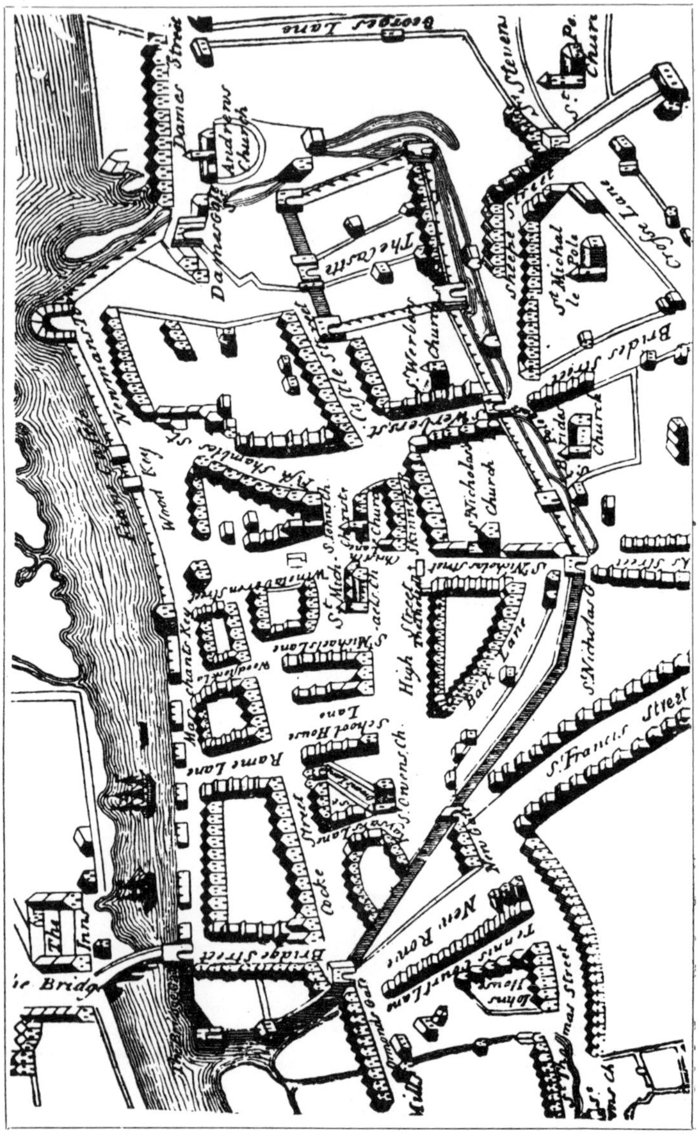

Plan of Dublin showing Castle Mills in 1601.

Hist. Dublin, Harris, 1766.

Mary's was designated " del dam," to distinguish it from St. Mary's Abbey opposite, across the river; in the

I. THE CASTLE MILLS, DUBLIN.

same manner that the earliest chapel of Liverpool, built on the edge of the Mersey, and designated St. Mary del Key, was distinguished from the mother church of St. Mary, Walton. The mill-dam at Dublin was there before 1179, when a deed attested by Goodmund, priest of Sainte Marie del Dam, shows it to be then in existence beside the church. About the year 1195 Prince John granted Doubleday the site for his mill on the ancient mill-stream; not long afterwards the monks of the abbey acquired the mill; and ultimately, in 1248, Henry III. established his Castle Mills on the same race, which then and down to modern times was the best spot within the walls of Dublin for working a watermill.

1. Site, &c., 1248.

Hist. Dublin, Gilbert, 1859, ii. 1.

The corporation of Dublin are early found in possession of the new mills as leaseholders under the Crown; a rental of Dublin city for about the year 1290 containing the memorandum that "the City used to render for the farm of the mills juxta the Castle 13 marks [£8 13s. 4d.] per annum; but the said mills are now in the hand of the lord the king." There is no actual record of this lease extant, apparently; but if it were in existence in the time of John la Warre, whose name as bailiff occurs in the same rental, it would be about the year 1245. Ere long, in any case, the leasing-out system was discontinued in favour of direct working on behalf of the Crown.

Hist. and Mun. Doc. Ireland, 214.

2. Very soon after the building of the Castle Mills a difficulty arose; the monks of St. Mary's Abbey complaining that their district had been invaded, so that they were losing trade to the value of 10 marks per annum, and setting forth their grievance in a petition to the king. Henry, not unmindful of the civil law of soke rights, ordered an enquiry to be held in Dublin; and it being then proved that matters were as the monks alleged, they were granted

2. Compensation to St. Mary's, 1254.

I. THE CASTLE MILLS, DUBLIN.

2. Compensation to St. Mary's, 1254.

in compensation a plot of land with an annual rental equal to their loss on milling.* The title of the monks to the plot was questioned some years later (1319) by the Exchequer; but the monks again appealed to the king; and the rolls being searched, the original writ of Henry III., as quoted below, was found in the Exchequer. On this occasion a record was made in the Dublin Exchequer, comprising the various facts, and giving a succinct and interesting account of the royal deference to ordinary manorial milling law.

Mem. Rolls of Ireland, 13 Ed. II.

Edwardus &c. thesaurario et camerariis suis de Scaccario Dublin salutem :—

Ex parte dilecti nobis in Christo abbatis Beate Marie juxta Dublin nobis est ostensum quod cum ipse per quoddam breve, dudum per consilium domini Henrici quondam regis Anglie avi nostri terre predicte factum et sigillo ejusdem avi nostri quo tunc utebatur in eadem terra, sub testimonio Johannis Filii Galfridi tunc justiciari ipsius avi nostri terre ejusdem consignatum tenuit ipse, quam et predecessores sui, abbates ejusdem domus semper hactenus a tempore confeccionis brevis predicti pacifice tenuerunt unam carucatam et quatuordecim acras terre cum pertinenciis in Novo Castro in comitatu Dublin, in recompensacionem dampni quod quidam predecessor

Potsdam Palace Mill.

* Apropos to the foregoing is the modern story of Frederick the Great, when building his palace at Potsdam, respecting the objection a local miller entertained to having his windmill pulled down. The miller, whose forefathers had held the place, refusing to sell, it was ultimately allowed to remain; and it may still be seen in a restored condition and among ornate surroundings in close proximity to the palace. A pendant to this story is one told us regarding Dr. Hughes, late Bishop of Bangor, who was advised, on taking possession of the episcopal residence, to have removed an old watermill which was considered to impede the view of the landscape. "No, let it be," said the bishop; "I am the son of a miller, and the place reminds me of old times."

2. Compensation to St. Mary's, 1254.

suus, tunc abbas loci predicti, sustinerat, occasione construccionis molendinorum nostrorum juxta castrum nostrum Dublin, tunc de novo levatorum, prout per quandam inquisicionem, ad mandatum dicti avi nostri, inde factam et in custodia vestra, ut dicitur, existentem, plenius apparet:

Vos prefate thesaurarie et barones nostri de Scaccario predicto terram predictam pro eo quod breve predictum per nos seu progenitores nostros reges Anglie hucuesque non extitit nec existit per breve sub sigillo nostro Anglie affirmatum jam de novo cepistis in manum nostrum et ea, sic in manu nostra, tenetis in ipsius abbatis dispendium non modicum et jacturam: Super quo dictus abbas nobis supplicavit ut sibi remedium congruum facere velimus:

Nos super tenore brevis et inquisicionis predictorum certiorari et eidem abbati quod justum fuerit fieri volentes in hac parte vobis mandamus quod scrutatis rotulis indenturis et memorandis dicti avi nostri ac omnibus aliis nejocium illud tangentibus de tempore predicto que sunt in thesauraria nostra sub custodia vestra, ut dicitur, id quod inde invenire contigerit unacum tenore brevis et inquisicionis predictorum, nobis sub sigillo Scaccarii predicti, distincte et aperte mittatis et hoc breve. Ita quod ea habeamus a die Pasche in quindecim dies ubicunque tunc fuerimus in Anglia ut nos, inde certiorati, ulterius fieri faciamus quod tunc de consilio nostro duxerimus ordinandum. Et terram predictam cum omnibus inde perceptis si eam ea occasione et non alia in manu nostra ceperitis, eidem abbati per sufficientem securitatem interim liberari faciatis tenendam, ita quod de exitibus inde provenientibus nobis respondeat si ad nos debeat pertinere.

Teste, meipso, apud Eboracum duodecimo die Octobris anno regni nostri terciodecimo.

Venit hic tercio decimo die Octobris anno predicto.

Memorandum quod postea tercio die Marcii proximo sequenti prefati thesaurarius et camerarii unacum baronibus de Scaccario predicto processerunt ad retornum brevis predicti faciendum et mittendum domino regi in Anglia in forma que sequitur:—

Pretextu istius mandati Scrutatis rotulis indenturis [&c.] prefati thesaurarius et camerarii domino regi mittunt secundum tenorem hujus mandati sicut plenius patet in cedula huic mandato consuta.

Preterea thesaurarius et barones de Scaccario predicto pretextu hujus mandati terram in mandato isto contentam et nominatam occasione in eodem mandato contenta, et non alia, in manum domini regis nuper captam unacum omnibus inde perceptis abbati Beati Marie juxta Dublin per securitatem sufficientem liberari fecerunt tenendam secundum tenorem hujus mandati ita quod ipse de exitibus inde provenientibus domino regi respondeat, si ad ipsum dominum regem debeant pertinere, prout in mandato isto plenius continetur. Et sciendum quod ad hoc faciendum predictus abbas per fratrem Ricardum Mannyng commonachum suum, cantorem ejusdem domus, de precepto thesaurarii et baronum hic &c. per Ricardum de Eboraco clericum hujus Scaccarii

I. THE CASTLE MILLS, DUBLIN.

ejusdem abbatis attornatum receptum in hac parte sicut alibi patet inferius in hiis rotulis memorandum de hoc termino Sancti Michaelis &c. domino regi in Scaccario hic &c. securitatem invenit subscriptam videlicet:—Philippum de Somerton Willielmum de Asshebourne Johannem Rowe Willielmum le Carpenter Ricardum le Carpenter et Willielmum Ballard ad respondendum &c. Et super hoc iidem Philippus Willielmus Johannes Willielmus Ricardus et Willielmus presentes in curia hic &c. inde devenerunt plegii et manucaptores ejusdem abbatis in forma supradicta, conjunctim et diversim.

2. Compensation to St. Mary's, 1254.

Tenor cedule, mandato predicto consute, patet in sequenti:—

[A] In rotulis memorandum de anno regni regis Henrici tricesimo octavo [secundo?]:—Henricus &c. dilecto et fideli suo Johanni Filio Galfridi justiciario suo Hiberni salutem. Quia accepimus per inquisicionem quam a vobis fieri precipimus, in Hibernia, quod abbas et monachi Sancte Marie juxta Dublin dampnificantur singulis annis per molendina nostra de novo levata juxta castrum nostrum Dublin usque ad summam decem marcarum, volumus et vobis mandamus quod in recompensacionem hujusmodi dampni ecclesie predicte et pro salute anime nostre assignetis eisdem, ubi siccius et melius providere poteritis ad minus dampnum nostrum, terram sive redditum annuum ad valenciam decem marcarum per annum; nisi predecessores nostri aut nos ipsi priusquam iidem monachi molendina ibidem habere consueverimus. Et cum hujusmodi terram sive redditum eis assignaveritis nobis significetis cujusmodi et quam terram sive redditum eis assignaveritis, ut terram illam sive redditum eis confirmare possimus carta nostra.

Teste, meipso, apud Marleberge tercio die Junii anno regni nostri tricesimo secundo [1248].

[B] Item in rotulis memorandum Henrici regis &c:—Memorandum quod ad Pascha anno regni regis Henrici tricesimo tercio [1249] primo recepit abbas Sancte Marie quinque marcas &c. de Novo Castro inde dicte quinque marce debent subtrahi de sexaginta libris quas Warynus de Fisshacr debet pro firma ejusdem Novi Castri de dicto termino Pasche. Et sic in anno in annum et de termino in terminum.

[C] Item in rotulis memorandum de anno regni regis predicti tricesimo quinto [1251]. Monachi Sancte Marie juxta Dublin habent unam carucatam terre de dono domini regis in tenementis de Novo Castro, que vocatur Colygtheny, de qua terra percipiunt annuatim decem marcas et dominus rex octo denarios tantum: Quam terram dominus rex assignavit eisdem pro dampnis sibi illatis per molendina domini regis juxta castrum Dublin de novo levata.

[D] In magnis rotulis de anno regni regis Henrici tricesimo octavo [1254].

Henricus &c. thesaurario et camerariis Scaccarii sui Dublin salutem:—

Quia accepimus per inquisicionem quam fieri fecimus quod abbas et monachi Sancte Marie Dublin dampnificantur singulis

2. Compensation to St. Mary's, 1254.

annis per molendina nostra de novo levata juxta castrum nostrum Dublin usque ad summam decem marcarum concessimus eisdem abbati et monachis in recompensacionem ejusdem dampni unam carucatam et quatuordecim acras terre cum pertinenciis in tenemento nostro de Novo Castro quequidem terra appellatur Collgtheny, salvis nobis de predicta terra octo denariis per annum. Et ideo vobis mandamus quod eosdem abbatem et monachos terram illam tenere et pacifice possidere permittatis sicut predictum est quousque eandem terram carta nostra eisdem confirmaverimus.

Teste Johanne Filio Galfridi justiciario nostro Hibernie apud Dublin nono die Maii anno regni nostri tricesimo octavo [1254].

Et sciendum quod predictum breve, indorsatum in forma supradicta, unacum predicta cedula eidem brevi consuta, supradicto tercio die Marcii anno decimo tertio liberatum fuit predicto fratri Ricardo Mannyng sub sigillo hujus Scaccarii, ad deferendum in Anglium.

Edward to his Treasurer and chamberlains of the Exchequer of Dublin, greeting :—

Whereas, on behalf of our beloved in Christ the abbot of the Blessed Mary juxta Dublin, is shown to us that he by a certain writ heretofore made and given to him by the council of the Lord Henry, former King of England, our ancestor, under the seal our said ancestor then used, by the testimony of John Fitz Galfred, then justiciary of our said ancestor ;—that he and his predecessors, abbots of the said house, have ever hitherto, since the issue of the said writ, peaceably held 1 carucate and 14 acres of land, with appurtenances, in New Castle, county Dublin, in recompense for damage which a predecessor of his, abbot of the said place, sustained by reason of the erection of our mills juxta our Castle of Dublin, then newly built ; as fully appears according to the record of a certain inquisition held by mandate of our said ancestor, existing in your custody as is said :

And whereas you our said Treasurer and barons of Exchequer —on the ground that such writ issued by our progenitors, former Kings of England, does not remain, and no confirmatory writ by us exists under our Seal of England—now take the said land anew into our hand and hold the same, as in our hand, to the no slight loss and damage of the said abbot ; Wherefore the said abbot petitions us to be pleased to make him suitable remedy :

We, acting upon the tenor of the said writ and inquisition, and being willing that justice be done the said abbot in this matter, charge you to examine the rolls, indentures, and memoranda of our said ancestor and all other business records of the like of that period that are under your custody in our treasury as is said ; so that if it happen that matters are thence found to be according to the tenor of the said writ and inquisition, you shall specifically and expressly send the same to us, under the seal of the Exchequer, with this writ ; in order that we may have the same within the feast of Easter wherever in England we may be ; and that we may make such further

I. THE CASTLE MILLS, DUBLIN.

2. Compensation to St. Mary's, 1254.

order on the matter as by our council we may decide. If the said land, with all revenue arising from it, you have taken into our hand for the above reason and no other, deliver it into the holding in the meanwhile of the abbot under sufficient security, so that he may answer to us for its income if it should belong to us.

Witnessed by myself at York, October 12, in the thirteenth year of our reign [1319].

[The above writ] arrived here [Dublin Exchequer] October 13 in the above year [*sic*].

Memo. Subsequently on the 3rd of March next following the said Treasurer and councillors, with the barons of the Exchequer, proceeded to make reply to the above writ and send the same to the lord the king, as follows:—

In accordance with the writ *Examine the rolls, indentures, &c.*, the said Treasurer and councillors reply to the lord the king, according to the tenor of his mandate, as fully appears in the schedule stitched to such mandate.

Further, the Treasurer and barons of the Exchequer, in accordance with the said mandate, have delivered to the holding of the abbot of St. Mary's, Dublin, the land contained and named in such writ, lately for the reason stated in the writ and no other seized into the hand of the lord the king, with all receipts arising therefrom, under sufficient security, in order that the abbot may answer to the lord the king for such receipts if the land should belong to him, as is fully contained in his writ. And be it known that for this purpose the abbot, by brother Richard Manning, his fellow-monk, priest of the chantry of the said house, in accordance with a precept of the Treasurer and barons (through Richard of York, priest of this Exchequer, attorney for the said abbot), accepted on this behalf, as appears below in these rolls of Michaelmas term, &c., has found as security Philip of Somerton, William of Ashbourne, John Rowe, William the Carpenter, Richard the Carpenter, and William Ballard; who present in court here have separately and conjointly become pledges and bondsmen for the abbot in form above said.

The tenor of the schedule, stitched to the mandate above mentioned, appears in the following:—

[A] In the memorandum rolls of the thirty-eighth [thirty-second] year of the reign of King Henry:—Henry, &c., to his well-beloved and faithful John Fitz Galfred, his justice of Ireland, saluting. Whereas we learn by inquisition held in accordance with our instructions to you in Ireland that the abbot and monks of St. Mary's, near Dublin, may be endamaged every year, by our mill newly erected beside our Castle of Dublin, a sum reaching 10 marks, we will and direct that, in some recompense for the damage to the said church, and for the welfare of our soul, you assign to them a piece of land or an annual rental of the value of 10 marks per annum, as you may do so best, not lavishly, and with least loss to us; unless our predecessors or ourselves have been accustomed to have a mill there before the said monks. And when you shall have thus assigned them a land or a rental, signify the same to us, that we may be able to confirm the same by our charter.

2. Compensation to St. Mary's, 1254.

Witnessed by myself at Marlborough, June 3, in the thirty-second year of our reign [1248].

[B] Item in the rolls of King Henry:—Memo. At Easter in the thirty-third year of the reign of King Henry [1249] the abbot of St. Mary's first received 5 marks from New Castle: the said sum to be deducted [in the Exchequer receipts] from £60 which Warren de Fishacre should pay for the farm of New Castle for the said Easter term. And so from year to year and term to term.

[C] Item in the rolls of the thirty-fifth year of the said king:—Memo. That the monks of St. Mary's, near Dublin, have one carucate of land by gift of the lord the king, in our tenement of New Castle, called Colygtheny, from which land they receive annually 10 marks, and the lord the king eightpence; which land the lord the king has assigned to them in consideration of damage inflicted upon them by the mill of the lord the king newly erected beside Dublin Castle.

[D] In the Great Roll of the thirty-eighth year of the reign of King Henry [1254]:—

Henry, &c., to the Treasurer and his barons of Exchequer at Dublin, greeting.

Whereas we learn by inquisition held by our order that the abbot and monks of St. Mary's, Dublin, are endamaged every year by our newly built mills beside our Castle of Dublin to the amount of 10 marks, we concede to the said abbot and monks, in recompense for such loss, one carucate and fourteen acres of land, with appurtenances, in our holding at New Castle, which land is called Colygtheny; reserving to us from the said land eightpence per annum. Wherefore we order that you permit the said abbot and monks to peacefully hold and possess that land until we shall have confirmed the same to them by our charter.

Witnessed by John Fitz Galfred, our justiciary of Ireland, at Dublin, May 9, in the thirty-eighth year of our reign [1254].

And be it known that the said brief, endorsed in form above said, together with the said schedule stitched thereto, on the said 3rd day of March in the thirteenth year of the reign [of Edward II., 1319], was delivered to the said Richard Manning, under the seal of this Exchequer, for delivery in England.

3. The City and the Keepers, 1280.

3. Reverting now to the Castle Mills; in 1280 the burgesses of Dublin, or at all events those of them who would be compelled to grind there, appearing to have been discontented with the millers, arranged with the authorities at the Castle that the commonalty should be permitted to have a voice in the appointment of

I. THE CASTLE MILLS, DUBLIN.

the officials of the mills, the city paying for the right the sum of £20:—

3. The City and the Keepers, 1280.

Memorandum quod termino Sancti Hilarii anno regis Edwardi filii regis Henrici, decimo, maior et communitas civitatis Dublin fecerunt finem in Scaccario domini regis per viginti libras, ita quod nullus postea de civitate predicta admitteretur molendinis sub castro, ad firmam, nec in custodia, nisi gratis voluerint. Et hoc idem irrotulatur in Scaccario.

Memorandum. In Hilary Term, in the tenth year of the reign of King Edward, the son of King Henry, the mayor and commonalty of the city of Dublin have paid in the Exchequer of the lord the king a fine of £20, in order that in future no person of the said city may be admitted to the farm or custody of the mills below the Castle, unless they [the mayor and commonalty] shall freely have been willing. Let this be enrolled in the Exchequer.

4. Committal of Keeper, 1306.

Text, III. 126.

4. In due course the mills were placed in charge of a keeper, the duties of whose office have already been explained; but in 1306 this official had the misfortune to be accused of neglect or fraud and committed to the Castle:—

Memorandum quod compertum est per thesaurarium et barones quod Willielmus Molendinarius minus bene custodit molendina domini regis juxta castrum Dublin: et magni defectus inveniuntur in eisdem molendinis, ob defectum bene custodie quam predictus debit facere: in contemptum domini regis et damnum gravissimum. Ideo commissus est castro donec, &c.

Be it known that it is found by the Treasurer and barons [of the Exchequer] that William the Miller has kept with but little care the mills of the lord the king juxta the Castle of Dublin; and that a great defect has been found in the same mills, for the lack of that good custody which the said William should have rendered; in contempt of the lord the king and to his most serious loss. Therefore he is committed to the Castle until——

The incompleteness of this record imparts to the fate of the recalcitrant a somewhat uncanny aspect, in view of the severity of the punishment for unfaithful millers at the city mills, recited in the "Ancient Customs" of Dublin three years later (1309). However, in the present case, William would doubtless but be imprisoned for a period sufficient to impress upon him and others that the custody of a king's mill was not a trifling affair to be lightly disregarded.

Ibid., III. 199.

I. THE CASTLE MILLS, DUBLIN.

5. Repairs, 1314.

5. In 1312 the custody of the establishment was officially under the direct supervision of Nicholas de Balscote, who is found to be Chancellor of the Irish Exchequer. To this dignitary would be responsible the millers both for profitable working and good management; while he would observe the maintenance of the soke, and would be entitled to hear cases of tenants absenting themselves from the mill or otherwise injuring its revenue. In this circumstance we may find the earliest direct instance of the so-called "court of the mills"—a popular medieval term, not an official title for any specially constituted court—of which we, perhaps, have the fullest illustration at Dee Mills, Chester. Balscote, no doubt, engaged practical keepers, who would be sworn to good and faithful service, as in the case shortly to be recounted. Between December 1312 and March 1314 he put the mills in thorough repair and purchased a pair of Welsh stones; and on the latter date presented to the Irish Treasury his bill of expenses, for which Edward II. issued the subjoined writ for payment:— Text, IV. ch. III.

Mem. Rolls of Ireland, 7 Ed. II.

Edwardus Dei Gratia, &c. Thesaurario et baronibus suis de Scaccario Dublin salutem.

Allocate Nicholi de Balscote, custode molendinorum nostrorum juxta castrum Dublin, super compotum suum ad Scaccarium predictum redditum :—

Quinquaginta et tres solidos et duos denarios quos posuit et expendidet in mæremio, lignis, bordis, clavis, et in aliis necessariis, emptis pro rotis dictorum molendinorum de novo constructis et aliis diversis defectibus in eisdem molendinis reparatis, a decimo octavo die Decembris anno regni nostri quinto usque ad primum diem Marcii anno regni nostri septimo; sicut continetur in rotulis de particularibus.

Et quadraginta solidos et sex denarios quos posuit et expendidit in stipendio diversorum carpentariorum, et pro reparacione: per idem tempus sicut continetur ibidem.

Et viginti et quinque solidos sex denarios et obolum quos posuit et expendidit in ferro et calabe, emptis pro fusilles et aliis grossis et minutis instrumentis ad dicta molendina necessariis de novo constructis: necnon in stipendiis cujusdam fabri, predicta instrumenta de novo construentis, et eciam sustinentis et acuentis, quociens necesse fuit: per idem tempus sicut continetur ibidem.

I. THE CASTLE MILLS, DUBLIN.

Et viginti et octo solidos et novem denarios quos posuit et expendidit in uno molari Wallensi empto ad predicta molendina: necnon in cariagio et reparacione ejusdem antequam loco suo ponebatur: per idem tempus sicut continetur ibidem.

5. Repairs, 1314.

Et quatuor solidos et novem denarios quos posuit et expendidit in sepo et uncto pro sustenacione dictorum molendinorum, quociens necesse fuit: per idem tempus sicut continetur ibidem.

Et sexaginta solidos septem denarios et obolum quos posuit et expendidit in sustenacione ejusdem equi dictis molendinis deservientis: necnon et pro ferrura ejusdem.

Nisi pecuniam illam ei prius allocaveritis per aliud breve nostro.

Edward the King, &c. To the Treasurer and barons of his Exchequer at Dublin, greeting.

Award to Nicholas de Balscote, custodian of our mills near the Castle of Dublin, upon his account rendered to the said Exchequer:—

53s. 2d. which he expended in structural timber, boards, keys [or nails], and other necessaries purchased for the wheels of the said mills, newly constructed; as well as for the repair of divers other defects in the said mills—from December 18 in the fifth year of our reign to March 1 in the seventh year of our reign; as set forth in the roll of particulars:

40s. 6d. which he expended in the wages of various carpenters executing the repairs during the same period; as stated in the same roll:

25s. 6½d. which he expended in iron and in sharp points bought for picks, and other large and small necessary tools made for the restoration of the mills; as well as in the wages of smiths making and sharpening and keeping in order the said tools as required; within the same period:

28s. 9d. which he expended in one Welsh millstone bought for the said mill; and in the carriage and dressing of the same before it was fixed in its place; within the same period:

4s. 9d. which he expended in soap and oil for the use of the said mills as occasion required; within the same period:

60s. 7½d. expended in the keep of the horse serving the said mill and in ironwork for the same:

Unless the said amount you may have already awarded him by any other warrant of ours.

6. Damages during War, 1316–17.

6. In March 1314 the mills were entrusted to two custodians (evidently lessees), Walter de Keppok and Randolph de Loung, who retained possession till March 1319. At this date, upon their giving up possession to two sworn keepers, their accounts were audited by the Exchequer, and it was then alleged against them that they had not accounted for £17 18s. due on various items during the period of five years. In

6. Damages during War, 1316–17.

reply they petitioned the Court, explaining that these alleged arrears should not be debited against them. This was the period of the struggle with Bruce on Irish soil, and the Castle millers, it appeared, had fared badly. They had ground grain for various dignitaries of the Castle—a couple of viceroys, a treasurer, and a justiciary; and these had not paid the toll due. Two of the horses serving the mills, which were missing and with which they were debited, had been seized for military purposes. One of the city gates already mentioned, the Gate del Dam, close to the mills, had been blocked up for an entire year, (the belfry of the adjoining church, Beate Marie del Dam castro nostro predicto contiguum, being taken down to find stones for the blockade); so that people on that side of the city could not bring grain to one of the mills, and took it elsewhere, as by law they were entitled to do; all that the mill earned during the year being 35s. While in the next year the same mill was broken down and lay prostrate, so that all it casually earned during that year was 40s. Thus for neither of these years could they pay the full rent of the mill—10 marks. These drawbacks together, they alleged, made up the £17 18s. now said to be due from them. The Exchequer thereupon ordered the mayor and commonalty of Dublin to convene a jury of twelve citizens, after Easter, to try the case—this being a veritable "Court of the Mills." The Exchequer ordered that the jurymen were not to be related to the millers; on the other hand, the mayor acted very fairly in summoning, as will be seen from the list, fourteen ordinary citizens, and six pistors who would be able to bring practical knowledge to bear on the issue. The jury found in favour of the defendants, and in the end (1320) it was found that they only owed the Court twenty pence.

Close Rolls of Eng., 12 Ed. II., m. 1.

Text, III. 219.

Ibid., ante, p. 13.

I.
THE CASTLE MILLS, DUBLIN.

6. Damages during War, 1316–17.

Memorandum quod Walterus de Keppok et Radulphus de Loung, nuper custodes molendinorum domini regis juxta castrum Dublin, compotum suum ad Scaccarium hic reddidissent de tempore quo ipsi custodiam habuerunt molendinorum predictorum: super quemquidem compotum ipsi in arreragiis remanserunt de decem et septem libris, decem et octo solidis; prout plenius patet per inspectionem compoti ejusdem.

Postea venerunt hic, coram thesaurario et baronibus, prefati Walterus de Keppok et Radulphus de Loung; et per quamdem peticionem suam in Scaccario hic inde exhibitam graviter conquesti fuerunt quod ipsi super predictum compotum suum indebite onerantur de tolneto octodecim crannocorum frumenti de blado, videlicet, Edmundi de Botiller, nuper justiciarii Hibernie: Et sexaginta crannocorum et quinque pecks frumenti de blado Johannis de Hothum, nuper domini regis locum tenentis in Hibernia: Et sexaginta crannocorum frumenti de blado Rogeri de Mortuo-Mari, similiter locum domini regis nuper tenentis ibidem: Et decem crannocorum frumenti et brasei de bladis Magistri Walteri de Istlep, thesaurarii Hibernie—quequidem blada, tempore pro predicti Walterus de Keppok et Radulphus de Loung custodiam molendinorem predictorum habuerunt, per vices diversas ibidem molebantur, et unde ipsi nullum ceperunt seu habuerunt tolnetum: et sic ipsi per surrepcionem curie hic de tolneto bladorum predictorum, quod ad decem crannocos et amplius se extendere debuit, minus juste in compoto suo predicto versus dominem regem onerantur, ut dicunt.

Queritur etiam quod ubi homines Arnaldi le Poer ceperunt ad eisdem custodibus molendinorum quendam equum qui deserviebat molendis predictis, precii quatuordecim solidorum. Et homines similiter prefati Rogeri de Mortuo-Mari postea ceperunt de eisdem custodibus quendam alium equum, emptum pro molendinis predictis, precii novem solidorum. Iidem custodes molendinorum nichilominus inde onerati sunt versus prefatum dominum regem in compoto suo predicto.

Dicunt eciam quod in eodem compoto suo ipsi indebite onerantur de decem marcis de exitibus: videlicet, unius molendinorum predictorum per unum annum integrum, ubi ipsi de exitibus ejusdem molendini in eodem anno non plus receperunt quam treginta et quinque solidis, occasione, videlicet, obstructionis Porte del Dam que toto eodem anno et amplius clausa fuit et obstructa ob guerram communem per Scotos et Hibernicos in hac terra tunc habitam. Ita quod nullum cariagium bladi per portam illam ad predictum molendinum per idem tempus fieri potuit.

Item queriter quod ipsi, in predicto compoto suo, ultra modum onerantur de decem marcis de exitibus ejusdem molendini pro anno proximo sequenti, ubi idem molendinum confractum fuit et per decem et octo septimanas prostratum jacuit, et nullum proficuum per idem tempus inde percipi potuit, nisi tantummodo quadraginta solidi. Et sic ipsi de predictis decem marcis pro eodem molendino quoad illum annum plenarie respondere non potuerunt, nec de aliquo proficio ultra predictos quadraginta solidos, in forma predicta inde perceptos.

I. THE CASTLE MILLS, DUBLIN.

6. Damages during War, 1316–17.

Unde supplicarunt, &c.

Et thesaurarius et barones concordarunt et ordinaverunt coram eis inquisicionem per sacramentum proborum et legalium hominum de civitate Dublin; et qui predictos Walterum de Keppok et Radulphum de Loung nulla affinitate. Et preceptum est maiori et ballivis civitatis Dublin quod venire faciant hic, &c., die Mercurii proxima post tres septimanas Pasche duodecim, &c.

Ad quem diem venerunt predicti Walterus et Radulphus et juratores similiter videlicet:—

Lucas de Hinkeleye	Nicholaus le Brewer
Thomas le Skynner	Phillippus de Bonevile
Johannes Bride *pistor*	Geraldus le Young
Alexr Boluyng	Ricardus de Swerdes
Johannes Siward *pistor*	Ricardus de Wenleie *pistor*
Galfridus Wallensis	Walterus le Lang
Johannes de Stradbaly	Walterus le Porter
Walterus de Bredon *pistor*	Johannes le Porter
Walterus de Valle	Andreas Rowe et
Stephanus Curtois *pistor*	Johannes Jordan *pistor*

Qui juratores dicunt . . . decem et octo crannocis frumenti, &c. Recipisse debuerunt novem crannocos et duos pecks bladi et brasei: precium cujuslibet crannoci dimidia marca; unde summa pecunie se extendit ad sexaginta et unum solidos et octo denarios, ut de tolneto, videlicet, inde debito, ubi nichil inde perciperunt, nec aliquid inde habere potuerunt.

Dicunt eciam quod homines Arnaldi et Rogeri, predictos duos equos, precium unius quatuordecim solidi et precium alterius novem solidi, vi ceperunt de predictis custodibus molendinorum, et eos abduxerunt contra eorundum custodum voluntatem. Dicunt eciam, &c.

Pretextu cujus veredicti prefati thesaurarius et barones, ex equitate curie, processerunt ad allocandum predictis Waltero et Radulpho in arreragiis suis predictis sexaginta et unum solidos octo denarios pro predicto tolneto, proveniente de predictis centum quadraginta et octo crannocis et quinque pecks, tam frumente quam brasei de bladis predictorum E. le Botiller, J. de Hothum, R. de Mortuo-Mari et W. de Istlep; pro eo quod eadem blada et braseum ibidem molebantur per predictum tempus absque nullo tolneto inde prestando; prout compertum est inquisicionem predictam.

Et similiter allocaverunt eisdem Waltero et Radulpho viginti et tres solidos de precio predictorum duorem equorum.

Item allocaverunt eisdem quator libras decem et octo solidos quator denarios de exitibus molendini predicti, de anno regni regis nunc nono, pro eo quod iidem custodes molendinorum de exitibus ejusdem molendini percipere non potuerunt illo anno amplius quam triginta et quinque solidos, pro eo quod toto illo anno Porta del Dam obstructa.

Item allocaverunt eisdem custodibus molendinorum quatuor libras tresdecim solidos et quatuor denarios de exitibus molendini predicti de supradicto anno decimo [regni regis] pro eo quod iidem custodes de exitibus ejusdem molendini amplius percipere non

I. THE CASTLE MILLS, DUBLIN.

potuerunt quam quadraginta solidis, pro eo quod idem molendinum in illo anno per decem et octo septimanas confractum fuit et prostratum jacuit antequam reparatum fuit.

6. Damages during War, 1316–17.

Unde summa pecunie, juxta veredictum inquisicionis predicte, prefatis custodibus molendinorum in hac parte allocate, se extendit ad tresdecim libras decem et sex solidos quatuor denarios, unde ipsi custodes exoneracionem consecuti sunt in arreragiis suis pre dictis: sicut plenius continetur in compoto eorundem custodum in magnis, videlicet, rotulis de hoc anno decimotercio.

Et sic sunt adhuc super eosdem custodes quatuor libre viginti denarii, unde ipsi postea solverunt quatuor libras, sicut patet per quandem talliam de Scaccario hic que eis inde allocantur in compote suo predicto. Et sic ipsi adhuc inde debent viginti denarios.

Memorandum. Walter de Keppok and Randolph de Loung, late keepers of the mills of the lord the king juxta the Castle of Dublin, should have rendered to the Exchequer here their account for the time they had the said mills: upon which account they have remained in arrears amounting to £17 18s., as fully appears from an inspection of their account.

Subsequently the said Walter and Randolph appeared before the Treasurer and barons here, and by a certain petition presented to this Exchequer made serious complaint that on the said account they are unduly charged with toll as follows:—On 18 crannocs * of wheat flour for Edmund Butler, formerly Justiciary of Ireland; on 60 crannocs 5 pecks of wheat flour for John de Hotham, formerly Viceroy in Ireland; on 60 crannocs of wheat flour for Roger de Mortimer, also late Viceroy in Ireland; and on 10 crannocs of wheat flour and malt for Master William de Istlip, Treasurer of Ireland. This grain was ground at various times at the mills while Walter and Randolph were custodians there; but, as they say, they took no toll from it, nor have they had any; and thus, little by little, they have been unjustly charged in the account of the lord the king for the above toll, which should amount to 10 crannocs or more.

They also complained that the men of Arnold le Poer took from the same keepers a certain horse of the value of 14s. which served the said mills; and in the same way the men of Roger Mortimer aforesaid subsequently took from them a certain other horse bought

---

* The case throws some light upon the capacity of the crannoc. In the verdict of the jury 2 pecks are reckoned as a fourth of a crannoc, which therefore was equal to 8 pecks or 2 bushels.

Shuttleworth, Accounts, Ch. Soc., xliii. 558.

"The word 'crannoc' is not to be found in *Promptorium Parvulorum* (1449), Halliwell's *Dictionary of Archaic and Provincial Words*, Brockett's *Glossary*, &c.; and we have been unable to ascertain the quantity represented by the term."

Hist. and Mun. Doc. Ireland, xxxiv.

"The Irish crannoc was a basket of twigs lined with the skin of a beast, and had no certain dimensions, but was understood to contain the produce of seven score sheaves of corn, which, however, must have varied greatly. Perhaps it generally amounted to about the quantity of the barrel of Bristol brought into use in Ireland when the English settled there: hence occurs the frequent use of the term 'crannoc' for 'barrel.'"

I. THE CASTLE MILLS, DUBLIN.

6. Damages during War, 1316–17.

for the said mills at the price of 9s. Nevertheless, the said keepers are debited with these values in the account of the lord the king.*

They also pleaded that they are unduly debited in the same account with 10 marks [£6 13s. 4d.] for multure—that is to say, the multure of one of the said mills during one entire year [1316]; whereas in that year the receipts received at that mill were no more than 35s., on account of the obstruction of the Gate del Dam, which for all that year and more was closed, owing to the war then prevailing in this country between the Scotch and the Irish, so that no corn could be conveyed to the mill by that gate during that time.

It was also complained that in the said account, contrary to custom, they are debited with 10 marks for multure at the same mill for the year next following [1317]; whereas the mill was broken and indeed laid prostrate for eighteen weeks, so that no profit during that time could be made out of it, except in one way or another 40s. Thus for the said 10 marks for the full year at the said mill they could not respond, nor for any profit beyond the said 40s. received there in the manner stated.

Wherefore they entreated [to be heard before a jury].

The Treasurer and barons agreed, and ordered the holding of an inquisition before them on oath by honest and legally qualified men of the city of Dublin, the same not being related to the said Walter and Randolph; and precept was issued to the mayor and bailiffs of the city to convene twelve men here on the Wednesday next after the three weeks of Easter.

On which day came Walter and Randolph and the jury—[*See original text*]. Which jurors [after reciting the details of the claim as above given] say that the said keepers should have received 9 crannocs 2 pecks wheat [flour] and malt, price half a mark per crannoc; the total sum amounting to 61s. and 8d. for toll—that is, the toll with which they are debited, although they received none of it and could not obtain it.

They also say that the men of Arnold and Roger carried off the two said horses, the value of one being 14s. and of the other 9s., from the said keepers against their will. They also say [that the keepers are entitled to the allowance claimed for loss of multure].

In accordance with the aforesaid verdict the Treasurer and barons by equity of the Court have given judgment and allotted the said Walter and Randolph as against their arrears aforesaid 61s. and 8d. for toll arising out of the grinding of 148 crannocs and 5 pecks of wheat flour and malt ground for Butler, Hotham, Mortimer, and Istlip; for that the said wheat and malt were ground for them there during the said period without any toll being received therefor, as

* These horses were of fair current value. In 1466 Dame Cecile, lady of Torbock, bequeathed "13s. 4d. to my son Thomas to buy him an horse." On the other hand, the tanner of Tamworth, hero of an adventure with Edward IV. (*Reliques of Ancient Poetry*), rode "a mare of four shillings." But at Dublin in 1315 John Fitz Thomas was allowed £10 for his war-horse, alleged to be worth £20, killed in the war with the Scots; and Robert Bagot, justiciary, was awarded £20 for another war-horse purchased by him for the king's use.

I. THE CASTLE MILLS, DUBLIN.

6. Damages during War, 1316–17.

found by the said inquisition. They have also allowed 23s., the value of the aforesaid two horses. Also they have awarded £4 18s. 4d. for multure of one mill during the ninth year of the reign of the present king [1316], because the keepers could not gain more than 35s. that year, on account of the Gate del Dam being obstructed. Also they have awarded £4 13s. 4d. for multure of the same mill in the tenth year of the reign of the king [1317], as in that year the keepers could gain no further multure than 40s., on account of the mill lying broken and prostrate for eighteen weeks before it was repaired.

Therefore the sum allotted to the keepers in this matter, in accordance with the aforesaid verdict of the inquisition, amounts to £13 16s. 4d., by which the keepers are consequently exonerated from the arrears aforesaid; as fully contained in the account of the said keepers entered on the great rolls for this thirteenth year [of the reign of the king, 1320].

Over and above the foregoing is also a sum of £4 20d. charged against the said custodians; but of this they have since paid £4, as appears by certain tallies of the Exchequer here, which is credited to them on their account. And thus they still owe in the matter 20d.

The sum and substance of the claim and award, if put in the form of a modern statement of account, would assume some such aspect as the following:—

DUBLIN CASTLE MILLS, MARCH 1319.

WALTER DE KEPPOK AND RANDOLPH DE LOUNG
IN ACCOUNT WITH THE EXCHEQUER OF IRELAND.

Cr.

By Toll unpaid on grinding to order of—				cr.	pks.
Edmund Butler, Justiciary				18	0
John de Hotham, Viceroy				60	5
Roger de Mortimer, Viceroy				60	0
William de Istlip, Treasurer				10	0
				148	5
$\frac{1}{16}$th toll on 148 cr. 5 pks. of wheat and malt				9	2

	£	s.	d.	£	s.	d.
By Value of 9 cr. 2 pks. wheat flour and malt @ ½ mark per cr.				3	1	8
,, ,, 1 mill-horse seized by R. de Mortimer				0	9	0
,, ,, 1 ,, ,, A. le Poer				0	14	0
,, Rental, already paid for 1316, of one mill, idle by the Gate del Dam being closed	6	13	4			
Less Casual Toll during the year	1	15	0			
				4	18	4
,, Rental already paid for 1317, the said mill being broken and prostrate	6	13	4			
Less Casual Toll during the year	2	0	0			
				4	13	4
				£13	16	4
,, Sundry Accounts (paid except 20 pence, as per tallies of the Exchequer)				£4	1	8

I. THE CASTLE MILLS, DUBLIN.

6. Damages during War, 1316–17.

The claim gives us a brief glimpse of the stirring times during which the two lessees held the Castle Mills. The entire English military power in Ireland was absorbed in encountering Bruce, who, by the closing of the Gate del Dam and the injury of the Castle mill actually within the walls, seems to have been met at very close quarters. Roger Mortimer, Earl of March, had landed in Dublin in that year (1316), having been appointed to succeed Butler, and in a short time they were both in the field at the head of 30,000 men. John de Hotham, an ecclesiastic standing high in the confidence of Edward II., had taken charge of affairs at the Castle during the absence of Butler in 1315–16; and though designated as viceroy by the millers, he was but a locum tenens for that absent official, and his name does not occur in the published lists of those who have held that supreme office. Walter de Istlip, Treasurer, had left Ireland in 1317 to quell riots in Lancashire; but returned in 1319, and is then found issuing a writ to the Mayor of Dublin regarding an old mill in the city, as mentioned later. Arnold le Poer, who had taken the field with Butler in 1316, became famous six years later in his contest with the Bishop of Ossory, on the occasion of Lady Kytiler being arraigned before his lordship for witchcraft. "Go to your church with your decretals and preach there," quoth the indignant Arnold to the prelate; but the would-be reformer ere long died in prison.

7. Oath of Keepers, 1319

7. Among the jurymen summoned to try the foregoing case at Easter, 1320, were John Siward and Galfred of Wales, these being two of the new sworn keepers now installed in the Castle Mills in place of the lessees. These keepers, eight in all, were sworn in on April 8, 1319, and four days later were furnished with a copy of an inventory of the plant of the mills

I. THE CASTLE MILLS, DUBLIN.

7. Oath of Keepers, 1319.

at the time of their entry. The eight men comprised the entire staff of the mills; and every one of them, whether of high or low degree, was personally held accountable for the proper and honest management of the establishment. A similar instance of this separate and conjoint responsibility occurs at Chester in the next century, when not only the official keepers but the general servants at Dee Mills were included in an indictment for extortion. We have also a continental instance comprising the swearing in of an entire staff:—

Text, IV., ch. III.

Stâtuta Massil., lib. i., cap. 55.

De molneriis & molnaironis & mulateriis & duobus officialibus super hoc eligendis—predicti tam molneri quam molnaironii jurant ad sancta Dei Evangelia bene et fideliter tam bladum quam farinam custodire & fideliter molere bladum.

Of millers and mill-servants and muleteers and two officials in this matter chosen—the aforesaid, as well the millers as the servants, swear on the Holy Gospels to well and truly take charge of the grain and flour, and grind the corn faithfully.

In the case of the Dublin keepers the form of the oath is less fully given than in the similar case at Chester, but its import is quite clear:—

Memorandum quod octavo die Aprilis anno predicto venerunt hic Henricus Wircestr, Thomas Turmyn, Robertus de Tristeldermot, Galfridus Wallensis, Willielmus Devensis, Walterus Brown, Robertus Russell et Johannes Syward et prestiterunt domino regi sacramentum suum coram thesaurario et baronibus hic, quod ipsi bene et fideliter domino regi et populo suo deservient in custodia molendinorum ipsius domini regis juxta castrum Dublin; et quod de exitibus molendinorum eorundem eidem domino, regi, de anno in annum ad Scaccarium hic bene et fideliter respondebunt. Et super hoc eodem die facta fuit quedam dividenda de rebus inventis in molendinis predictis et prefatis custodibus molendinorum predictorum liberatis: cujus tenor patet in sequenti:—

Dividenda facta de vasis et ceteris utensilibus inventis in molendinis domini regis sub castro Dublin, octavo die Aprilis, &c., et liberatis Henrico de Wyncestr, Thome Tormyn, Roberto de Tristeldermot, Galfrido Wallensi, Willielmo le Devensis et Roberto Russell, custodibus molendinorum predictorum; videlicet:—Sunt ibidem dua dolia apparata pro blado et farina imponendis et custodiendis. Item quator bille et j hope et dimidium parvi valoris, pro tollneto recipiendo; et unus tribulus ad idem.

Et sciendum quod una pars dividende predicte liberata fuit

predictis pistoribus, duodecimo die Aprilis, anno predicto; et alia pars ejusdem dividende remanet penes rememoratorum, videlicet in filacio de extentis et inquisicionibus de hoc anno [1319].

---

7. Oath of Keepers, 1319.

Memorandum. On the 8th of April, in the year above mentioned, came here Henry of Worcester, Thomas Turmyn, Robert of Tristeldermot, Galfred of Wales, William of Devon, Walter Brown, Robert Russell, and John Syward; and bound themselves by oath to the lord the king before the Treasurer and barons here that they will well and faithfully serve the lord the king and his people in the custodianship of the mills of the said lord the king beside Dublin Castle; and that from year to year they will well and truly answer to the said lord the king in the Court of Exchequer here for the revenues of the said mills. And thereupon the same day was made an inventory of the articles found in the aforesaid mills and delivered up by the previous keepers: its tenor being as follows:—

Inventory of Plant, 1319.

Inventory of dishes and such-like utensils found in the mills of the lord the king below the Castle of Dublin on the 8th of April, &c., and delivered to Henry of Winchester [previously termed as of Worcester], Thomas Tormyn, Robert of Tristeldermot, Galfred of Wales, William of Devon, and Robert Russell, keepers of the said mills; that is to say:—There are in the said place two tubs [or bins] prepared for containing in safe custody corn and flour; four bills [or millstone-picks], one hoop-measure, and a half hoop-measure of little value for the taking of toll; and one tribulus [or scoop-measure] for the same purpose.

And be it known that one part of the said inventory was given to the afore-named pistors on the 12th of April in the said year, and the other part remains among the records filed with the surveys and inquisitions of this year [1319].

The foregoing illustrates the subjection of keepers of royal mills solely to the supervision and, in the event of fraud, to the jurisdiction of the Exchequer; whilst keepers of any other mills in the city were subjected to the control of the corporation, as set forth in the civic confirmation in 1309 of certain ancient "good customs." The inventory of Dublin Castle Mills in 1319 will be noted to compare unfavourably with that of the more extensive mills of the king at Ardee in 1305. Among the items mentioned at Dublin were the hoop-measure and the tribulus, both of which occur in the finding of the jury at an interesting inquisition held at Chester about 1399; the first named being a measure and the second a toll-dish.

Text, III. 198

Ibid., III. 8.

Ibid., IV., ch. III.

I. THE CASTLE MILLS, DUBLIN.

8. The King and the City Mill, 1319.

8. At this period another Dublin watermill comes into the records. The Castle authorities seem just then to have been attempting to consolidate, in one or two of the usual methods, all possible soke rights of the king's mills. As already shown, it was in 1319 that the Exchequer questioned the title of St. Mary's Abbey to the Colygtheny plot of land, held as compensation for injury done by the Castle Mills; and it was in the same year that the same authorities took steps to ascertain what possible soke rights might be claimed in respect of the mill which now comes under notice. On July 19 the mayor and corporation were served with a writ, signed by Walter de Istlip, returnable for the next week, demanding payment of the rental of a mill near St. Werburgh's Gate for the past eighty-five years, at the rate of one pound of pepper per annum. This mill really had long been destroyed by the king's orders, a fact of which the authorities were doubtless well aware; the claim being made, therefore, solely to obtain from the corporation an acknowledgment of the title of the king to any milling rights that might appertain to the site. The following is the writ:—

Dublin White Bk., fol. 96.

Rex maiori et ballivus suis civitatis Dublin salutem. Quia constat nobis per inspectionem rotulorum memorandum de Scaccario nostro Dublin, quod tu ipse, maior, unacum communitate civitatis predicte, nobis teneris in quatuor viginti et quinque libris piperis, de arreragio redditus molendinorum aquaticorum extra portam Sancte Werburge juxta videlicet civitatem predictam, ab anno regni regis Henrici avi nostri quintodecimo usque ad annum regni nostri duodecimum de quibus in nullo nobis hucusque satisfactum est: Vobis precipimus, firmiter injungentes quod de bonis et catallis communitatis civitatis predicte, necnon et tui ipsius, predicte maior infra civitatem predictam, ad quorumcunque manus devenerunt, predictas quater viginti et quinque libras piperis, sine dilacione, levare faciatis, ita quod totum piper predictum promptum habeatis ad predictum Scaccarium nostrum nobis solvendis die Lune proxima post festum Sancte Margarete Virginis, et habeatis ibi tunc hoc breve.

Teste Waltero de Istlep, thesaurario nostro Hibernie, decimo nono die Julii anno regni nostri duodecimo

I.
THE CASTLE MILLS, DUBLIN.

8. The King and the City Mill, 1319.

The King to the mayor and bailiff of his city of Dublin, greeting. Whereas it is evident to us by inspection of the memorandum rolls of our Exchequer at Dublin that you, the mayor, with the commonalty of the city aforesaid, are indebted to us in 85 lbs. of pepper, being arrears of rent of the watermills outside the gate of St. Werburgh's * beside the said city, extending from the fifteenth year of the reign of our grandfather King Henry [1231] to the twelfth year of our reign [1319],† for which no manner of satisfaction has been given us; We send this precept, firmly enjoining you that upon the goods and cattle of the community of the said city, as well as of yourself the said mayor, you shall cause to be levied without delay the said 85 lbs. of pepper unto whosesoever hands in the said city they shall have come; in order that you produce the whole of the said pepper to pay us at our said Exchequer on the Monday next after the feast of St. Margaret the Virgin [July 20], and that you produce there and then this writ.

Witness Walter de Istlip, our Treasurer of Ireland. Dated July 19, in the twelfth year of our reign [1319].

In reply the city fathers pleaded, in the usual official Anglo-Norman, their unanswerable defence to the demand for rent, and incidentally made no claim to what milling rights might exist:—

Dublin White Bk., fol. 96.

Auant le temps qe sire Johan le fitz Geffrei vynt justice en Irlaund, e avant la venu sire Geffrei de Tureuill tresoror, un Rogier le coruiser de Bristuyt qe fut un aduentifs a la cite de Diuelyn, fist leuer un molyn desuth le chastel, pur le sist de quel molyn a cel temps e pur une venele auoir ouuerte vers la esglise Seint Piere de la Hulle pur carier blees vers le dit molyn a moudre, soleit meymes cel Rogier e autres tenauns de cel molyn porter un liure de pepure de rente par an a nostre seignur le roy. E puis cel temps les justices et les tressores qe virent cel molyn nusaunt et perilous au chastel et auxint trop pres as molyns nostre seignur le roy, fesent cel molyn abatre et del tut ouster. Et la place ou adonque fu le sist de cel molyn est ore un petit parce approprie au chastel, et un autre place de coste au adonqe fu lestang du molyn, ioust la terre qe fu a un Serlot de Ripon, qe rent chescun an a nostre seignur, le rey, dues south, les quex le conestable resceit. Et issint est nostre seignur le rey seisi du soyl dount cele rente dune liure de pepure soleit issir, e par ceste resoun ount les citezeins sea en arriere este descharges e semble a eux qe encore demoerent.

Before the time when Sir John Fitz Geoffrey became Justiciary in Ireland, and before Sir Geoffrey de Tureville, Treasurer, came

---

* Harris, the historian of Dublin (1766), says it is only "in latter times this gate has been called St. Wyburgh's Gate."

† If these dates are correct, the claim should have been for eighty-eight, not eighty-five, years' rental.

I. THE CASTLE MILLS, DUBLIN.

8. The King and the City Mill, 1319.

there, one Roger, the corviser [trumpeter ?] of Bristol, who was an adventurer [or newcomer] in the city of Dublin, caused to be erected a mill below the Castle; for the site of which mill at that time, and for having opened a road towards the church of St. Peter de la Hulle, for the carrying of grain to the said mill to be ground, the said Roger and other tenants of the mill used to be responsible for the payment of a pound of pepper per annum rent to our lord the king. But after that time the justices and treasurers, who saw the said mill annoying and perilous to the Castle and adjoining too nearly to the mills of our lord the king, caused the same to be pulled down and altogether abolished. And the place where formerly was the site of the mill is now a small park appropriated to the Castle; and another place alongside where formerly was the pool of the mill was the land of Serlot of Ripon, who rendered every year to our lord the king two sous, which the constable received. Thus our lord the king is seized of the soil whence the rent of one pound of pepper used to issue; and for that reason the citizens are discharged from liability for arrears, together with those who still dwell there.

The Sir John Fitz Geoffrey above mentioned was viceroy in 1245, and "Sir Geoffrey de Tureville" is apparently identical with Sir Geoffrey de Geneville, who held the same office in 1273. The mill is stated to have been built by Roger of Bristol therefore before 1245. Roger of Bristol was treasurer of St. Patrick's Cathedral 1219–23, and prebend of the church of St. Mary del Dam, near the Castle and the Castle Mills. The mill he built was not quite in this locality and was not one of the Castle Mills, being situated near St. Werburgh's Gate, on the upper part of the stream which drove the king's mills. Till its destruction by order of the king, as being too near the latter mills, the corporation had leased it, probably in the fee-farme of the city; and subsequently they claimed no title to rebuild it.*

Hist. and Mun. Doc. Ireland, 219.

* The corporation in 1261 made violent reprisals upon another watermill only mentioned in the records in connection with this circumstance. A dispute arose in that year between the corporation and the priory of St. John at Kilmainham with respect to the fishing-net of the former placed under the bridge of Kilmainham. The prior and his men having demolished this net, the corporation, adopting the same tactics, "went to a certain mill of the said prior, and, by force of arms, threw down the same."

I.
THE CASTLE
MILLS,
DUBLIN.

9. Modern.

9. There is little more of interest to record regarding the Castle Mills. The church of St. Mary del Dam lasted till the sixteenth century, when the residents in the Castle formed almost its only parishioners; the Gate del Dam and the Castle Mills, as kings' mills, enduring a century later. The latter, in the form of two little hursts on the stream to the east of the Castle near Dame Street, are shown (as well as other places mentioned in this account) in a plan of Dublin dated 1601. They were worked by the Crown till 1639, when they were granted by Charles I. to the corporation; being then known as Dame's Mills in Dame Street. Till the middle of the eighteenth century they yielded to the corporation a rental of £20 per annum, the dam producing £4 per annum.

Text, ante, p. 4

A RELIC OF SHREWSBURY ABBEY MILLS.

CHAPTER II.

SHREWSBURY ABBEY MILLS.

II. SHREWSBURY ABBEY MILLS.

1. Foundation, 1083.

1. THE history of the mills of Shrewsbury affords an excellent illustration of the manner in which, in the early days of the feudal system of the Conqueror, a milling soke was established and built up; as well as of the manner in which, later, despite fully confirmed rights, such a property was destroyed by, or in deference to, public agitation. It also illustrates, unfortunately, how scantily, incorrectly, and even unwillingly popular historians have bestowed their attention upon the uncongenial, yet not unimportant theme of corn milling.

The earliest actual title of Shrewsbury Abbey to the rights we have to consider is found in a charter of Roger Montgomery, Earl of Shrewsbury; but as the standard local historians, Owen and Blakeway, question the authenticity of this and other charters, we may, before quoting it, seek some confirmatory evidence of interest in earlier records. The historian Ordericus Vitalis, himself a Salopian and a monk, though not of this foundation, briefly recounts the circumstances of the establishment of Shrewsbury Abbey; a matter upon which much of the point of this sketch depends.

Cœnobium quoque novum ad orientalem portam propriæ urbis id est Scroburiæ in honore Sancti Petri principis apostolorum cœpit edificare, Rogerius comes scilicet, super fluminum qui Mola dicetur ibique in Sabrina delabitur. Eccl. Hist., 578, 581, 594.

Anno igitur ab incarnatione Domini mlxxxiij indictione quarta Rogerius comes, ut hortatus fidelis consiliarii commodos esse comprobavit, Guarinum vicecomitem et Picotum de Saia cæterosque

II. SHREWSBURY ABBEY MILLS.

1. Foundation, 1083.

proceres suos quinto Kal. Martii feria sexta convocavit. Manifesta vero causa et ad omnibus collaudata, cum eis ad ecclesiam B. Petri apostoli adiit, ibique se abbatium constructurum palam testibus multis devovit, totumque suburbium quod extra portam orientalem situm est, Sancto Petro donavit, et super aram per cirothecas suas donationem posuit.

Eodem anno Sagienses monachi Reinaldus et Frodo primitus illuc convenerunt et monachilis officinas Odelerio et Guarino aliisque multus adjuventibus condere cœperunt.

Primus abbas illius monasterii Gulielmo Rufo regnante eloquens Fulcheredus fuit.

A new monastery at the east gate of his own [baronial] town, Shrewsbury that is, in honour of St. Peter, principal of the apostles, he, Earl Roger, purposed erecting on the border of the stream called Mola, where it flows into the Severn.

Accordingly, in the year of the incarnation of our Lord one thousand and eighty-three, the fourth indiction,* Roger, encouraged by his faithful counsellors, that he might fitly verify his purpose, convened Warine the sheriff, Picote of Seez [in Normandy], and others of his nobles on the 5th–6th calends of March [March 3–4]. His object was made manifest to the applause of all when he went with them to the church of St. Peter the apostle, where publicly, in the presence of many witnesses, he vowed to construct the abbey, and gave to St. Peter's the whole of the suburb outside the east gate of the city; laying his gloves in testimony upon the altar.

The same year Reinaldus and Fredus, monks from Seez, first met there; and Odelerius and Guarin, with many others, able workmen, began to build the monastic offices.

The first abbot of the monastery was the eloquent Fulchard, of the reign of William Rufus.

The only items in the foregoing account now needing mention are such as will be found pertinent to this enquiry. The abbey was founded near the junction of the Mola, *i.e.* the mill-brook, and the Severn in 1083, and the erection of the monastery was commenced in the same year; the abbey being opened under the first abbot in the reign of William II., viz. in 1087. Odelerius, one of the monks who began the building operations, was the father of Ordericus, our historian.

The next source of information is Domesday. The

* Ordericus makes a clerical error in terming this the fourth indiction year, as that would be 1081; the year 1083 being the sixth indiction.

survey of Shropshire took place in 1086, and the II. SHREWSBURY ABBEY MILLS.
founding of the abbey, which was then in course of
erection, is twice referred to, and its grant of mills
recited :— 1. Foundation, 1083.

In Sciropesberie civitate facit Rogerius com̄ abbatiā 7 eid̄ ded̄ monasteriū S. Petri ubi erat parochia civitat̄ tantū de suis burḡibuz 7 molinis qd̄ xii lib̄ redd̄ monachis. Domesday, i. 253[a].

Dicunt anglig̃ burḡses de Sciropesberie . . . abbatie qua facit ibi comes dederit ipse xxxjx burḡses. Ibid., i. 252.

In the city of Shrewsbury Earl Roger is building an abbey, and has given to it the monastery of St. Peter's, where was a parish [church] of the city; and also as many of his burgages and mills as would yield £12 [per annum] to the monks.

The English burgesses of Shrewsbury say that the earl has given to the abbey he is building there 39 burgages.

The burgages were not on the abbey lands, but in the city; otherwise the citizens would have had nothing to say about them. To yield a total income of £12 per annum, Roger seems to have largely miscalculated the probable produce of the burgages and mills. The Domesday value of burgages in St. Alkmond's parish, (Ibid., i. 253.) in Shrewsbury, was 5d. each; and the average gablo payment for each of the fifty-two houses in (Ibid., 252[b].) Shrewsbury liable in the time of the Confessor had been 7½d. It is thus clear that the yield of the thirty-nine burgages would amount to very little, even with the value of their multure (which, of course, usually followed ownership) added thereto. The general income of the mills, therefore, was clearly relied upon to afford the main portion of the £12 per annum. But the value of mills in the entire county was only small, as hand-stones were doubtless in general use. There was one mill in the county, that of Stanton, which yielded 26s., and two others were rented at 20s. each; but (Text, II. 163.) these were high values, and the average was under

II. SHREWSBURY ABBEY MILLS.

1. Foundation, 1083.

5s. each; even 6d. per annum, as at Rodingtone, appearing in the schedule.* Hence the value of the (three) small rural mills which Roger gave to the abbey would be very far from making up the desired £12 per annum. Now, it is remarkable that the monks in no one of their future inspeximus charters, nor in their *History of the Foundation*, nor elsewhere, ever claim to have been endowed with these burgage rentals; and it will, indeed, be suggested, shortly, that, before the actual "foundation charter" was drawn up in 1087, this combined gift of city burgages and rural mills, given at the initial stage of the foundation in 1083, had been found, on experience, to be inadequate to meet Roger's ideas, and that he consequently made a suitable change in its nature when the abbey was opened and an abbot installed, in 1087.

2. Grant of City Multure, 1087.

2. Earl Roger's "foundation charter" is the next link in the evidences. From this we learn that, prior to and during the erection of the monastery, the earl and others had vowed various gifts to the foundation, but that the monks, who were trustees for the erection, had not been endowed with them. It is already apparent from Domesday that the abbey was being built in 1086; and we may gather from the later evidence of the *History of the Foundation* that it was in 1087 when the building was fairly completed, the abbey at length opened, and an abbot appointed. This, as Ordericus tells us, was Fulchard, of the reign of William Rufus. It was in the last quarter of 1087, therefore, that Roger's charter was drawn up. In it were formally recited all the gifts that had at different

Hist. Shby., 1825, ii. 10.

* Owen and Blakeway say: "The mill of Ellesmere forms a remarkable exception: it paid a rent of £10 in the time of the Confessor, which was doubled when Domesday was made." This is an error. No valuation is quoted for the mill: the £10 is the rent of the whole manor:—Ipsi com̄ ten̄ Ellesmeles: Ibi IIII hidæ, &c., &c: Ibi molin: T.R.E. reddeƀ x liƀ de firma: Modo xx liƀ.

Domesday, i. 254.

II. SHREWSBURY ABBEY MILLS.

2. Grant of City Multure, 1087.

Monast. Ang., 1682, i. 379[a] (Ex Reg. Abb.)

times been vowed; and the whole endowment was vested in the abbey and the abbot, who, the charter states, had already been appointed. Clearly the king William mentioned in the charter is William II.

Rogerus comes Salopesberiæ filiis suis et heredibus et baronibus et hominibus et omnibus amiciis suis Francis et Anglis salutem.

Sciatis me construxisse monasterium in suburbia civitatis Salopesberiæ in ecclesia sanctorum apostolorum Petri et Pauli quæ antiquitus ibi fundata erat, et concessisse eidem vicum illum totam qui dicitur Biforiete cum tribus molendinis et omnibus redditibus et tota terra quæ est inter duo flumina scilicet inter Sabrinam fluvium et Meolam, et ex alia parte fluminis Meole terrulam quandam quæ vocatur Aldefield.

Et quia monachi Sagiensis [Sayiensis] cœnobii de mea possessione erant in Normannia et boni testimonii viri, de illis quosdam accepi, et ad edificationem præfati monasterii constitui, nulla illis ejusdem loci donatione concessa, sed ut elemosinas meas, et aliorum fidelium susciperent et colligerent, et de beneficiis collatis constructionem præfati monasterii festinarent.

Postea vero per consilium regis Gulielmi et Roberti episcopi et uxoris meæ Adelaisiæ et filiorum meorum, unum illorum abbatem ibi posui, et locum illum liberum feci, nullo omnino alteri loco per subjectionem obnoxium. . . .

Dedi etiam eis domum in civitate ad sabulonem [tabulonem] ad opus ecclesiæ et moltam totius civitatis et curiæ meæ quando in civitate manerem. . . .

Quæ omnia ego eis concessi et sigilli mea auctoritate confirmavi.

Præcepi etiam filiis meis. ut locum illum diligerent et carum haberent. Et si forte contingeret ut ego in Anglia morirer ibi me ponerent et deinde fratribus loci illius necessaria providerent.

Testibus hiis—Roberto episcopo Cestriæ, Hugone et Roberto Belismo &c.

Roger, Earl of Shrewsbury, to his sons and heirs, his barons and people, and all friends, French and English, greeting.

Know ye that I have built a monastery in the suburb of the city of Shrewsbury, in the church of the holy apostles Peter and Paul, which was anciently founded there; and have conceded to the same all that road called Biforiete, with three mills and all [their] revenues, and all the land between the two rivers, that is to say, between the Severn and the Meole; also a certain land called Oldfield, on another part of the Meole.

And whereas the monks of the monastery of Seez, on my land in Normandy, were men of good repute, I chose certain of them, and appointed them for the erection of the said monastery. They were conceded nothing of the donations made to the abbey, but they collected and received gifts from myself and others of the faithful, and with these benefactions they hastened on the building of the said monastery.

II. SHREWSBURY ABBEY MILLS.

Subsequently, by the council of King William, Bishop Robert, my wife Adelaisa, and my sons, one of them [the monks] I have placed there as abbot;* and have made the abbey free, so that it shall not be obnoxiously subject to any other whatever.

2. Grant of City Multure, 1087.

I have given them also a house in the city for an office or registry for the use of the church [monastery];† and the multure of all the city and of my court when I reside in the city. . . .

The whole of which I have conceded and by the authority of my seal confirmed to them.

Also I have enjoined my sons to take care of that place and hold it dear. And if it happen that I should die in England they shall place me there, and shall provide thenceforth what may be necessary for the brethren of the place.

Witnesses to these—Robert, Bishop of Chester, Hugo and Robert Belismo [sons of the founder], and others.

From this document we gather how vivid was the interest of the founder in the monastery he had built and the abbey he had established at the old church of St. Peter, where he had laid his gloves on the altar. Since then he had at different times endowed it with many valuable gifts, had solemnly commended it to his sons' care, and had desired to be buried there; and there, on July 27, 1094, having died a monk of his own foundation, in the place of honour between the altars he was accordingly interred. The local historians question the genuineness of this charter; but finally admitting its "general unstudied correspondence with Domesday," accept it as evidence; and agree that "it must be considered the earl had granted to his monks the multure of the whole city."

Hist. Shby., ii. 10.

Round this grant all the following events directly centre. The thirty-nine burgages in the city

* The faulty phrasing of the original suggests that Roger appointed one of his sons to be abbot: an error into which, as several commentators point out without explanation, the antiquary Stowe has fallen. Owen and Blakeway omit this clause in their rendering of the charter (ii. 11), and elsewhere remark (ii. 106): "Stowe calls the first abbot the founder's son, but on what authority we cannot learn."

† "Sabulonem" (a gravel pit) is a misprint by Dugdale in *Monasticon* (1682), and the same appears in the edition of 1846. Owen and Blakeway reproduce this, and add, "The gift of a house for the purpose of digging gravel is not very intelligible" (*Hist.*, ii. 12). The obvious reading, however, is "tabulonem," a house for an office, a counting-house.

mentioned in Domesday are not claimed in the charter at all. Clearly, therefore, early in 1087, the earl had withdrawn them, as already supposed, and substituted the more efficient gift of the multure of the city, which is claimed. That is, he constituted the abbey mills the manorial mills of Shrewsbury, with the usual stipulation that they should grind for the lord's court when he should be in residence there. This was an ordinary service—free multure paid to the lord, in return for and in recognition of the enjoyment of the multure of the manor. The one was part and parcel of the other, and the whole story of the charter is thus compact and complete. Nothing, moreover, is more probable than that the earl—whose interest in the abbey's welfare was so strong, and whose faith in the monks' rectitude so great—should, on the one hand, give them what revenue was to be had from the city multure; and, on the other, entrust to them the duty of providing and maintaining, as all manorial soke-owners were bound to do, a sufficiency of adequate mills for the needs of the city. The three manorial mills thus to be maintained for the use of the public were all in the Abbey Foregate suburb of the town upon the Mola; or rather upon an ancient and long-extended mill-race headed up from the same near to the abbey.

2. Grant of City Multure, 1087.

3. Next follows a charter by "King William," whom Owen and Blakeway assume to have been the Conqueror, but whom all the evidences identify as William II.; the charter being granted apparently in July 1094, when Earl Roger, having become a monk of Shrewsbury, was an inmate of the abbey for but a few days before his death. It is not surprising, therefore, to find the local historians repudiating the authenticity of "this pretended charter." It might

3. An alleged Forged Charter, 1094.

II. SHREWSBURY ABBEY MILLS.

3. An alleged Forged Charter, 1094.

Text, post, p. 41. note.

probably be convenient to follow their example, and curtly dismiss it as a forgery; but while reasonable explanation of its terms is possible and tenable proof of forgery is lacking, it becomes a duty to accept and endeavour to somewhat explain it. They impugn the deed (which no doubt was drafted by the monks) on the following grounds:—

Hist. Shby., ii. 15, 16.

It relates how the earl, by the licence of King William, with whose name in the first person this pretended charter sets out [1], and Archbishop Lanfranc, and Peter, Bishop of Chester, who had died in 1085 [2], assembled monks there in the year from the incarnation of our Lord 1087, a mode of dating unknown in English charters of that age [3]. . . . The manner in which the Conqueror is made to speak of Earl Roger is perfectly ridiculous: that nobleman, after a lapse of years, was 'venerable' to the monks who shared his bounty, but it is absurd to suppose that a sovereign would thus style a subject [4].

In every one of these boldly advanced condemnations (numbered here, as above, for reference) the historians are utterly wrong, and if the document is to be proved a forgery it must be done on other grounds than those they advance.*

* [1] They fail to discover that this is a charter of William II., not of William I., and consequently that it does not set out in the first person with the name of the latter, but with the name of the former; under the auspices of whom and of Bishop Robert the abbey was opened in 1087.

[2] There appears to be neither error nor ambiguity in the allusion to Bishop Peter. It is true that he died in 1085, but it was he who in March 1083, as Ordericus states, had granted the original licence to Roger to establish the house. His authority did not die with him; and, as the charter states, in 1087 the monks ultimately "congregated," or assembled, under his original licence—not in his lifetime or presence, as the historians mistakenly conceive. Precisely the same statement as to Bishop Peter, with equal accuracy, is made in the *Historia*, quoted later.

[3] The charter is not dated in Arabic figures, "1087," in a mode then uncommon, but exactly as follows—anno ab Incarnatione Domini millesimo octogesimo septimo. Some one, later, has marked the figures "1087" in the margin as a reference (see *Monasticon*, edition of 1682), and the historians seem to have confused this with the charter itself.

Codex Dipl. Ævi. Sax., I. xxj passim.
Rymer's Fœdera, 5 Richard I. et seq.

[4] Sovereigns frequently and ordinarily styled subjects "Venerable." Instances may be adduced by the score, from the time of Ethelbald, who thus addressed Hubert (Ego' Aethibald . . . *uenerabile* Seruo dei Eauberhtæ), and other Saxon kings; to Richard I. in 1194 (Ricardus rex Anglia . . . *Venerabile* patri Hubert); Edward IV. in 1478 (Edwardus rex . . . *Venerabilis* pater Johannes); and so on. In fact, "Venerable" was an ordinary courtesy title usually applied to distinguished clerics: and it was in this capacity that, in 1094, the distinguished monk Earl Roger was so addressed. The same term, used in the admittedly genuine charter of Henry III. to Shrewsbury, quoted later, passes unnoticed by the local historians.

II. SHREWSBURY ABBEY MILLS.

3. An alleged Forged Charter, 1094. Monast. Ang., i. 380[b] (Pat. 1 Hy. VI., p. 5, m. 19).

Sanctorum patrum monemur exemplo possessiones quas nostris temporibus sanctæ ecclesiæ à fidelibus collatas noverimus, literarum memoriæ diligentur assignare, tum propter conferentum utilitatem quoniam eorum exemplo posteri ad bene agendum provocentur; tum propter malignorum perversitatem qui ad diripiendas possessiones ecclesiasticas toto conamine inhiare noscuntur. Nos itaque majorum imitantes industriam posterorum notitiæ scribendo transmittere curavimus quoddam pietatis opus quod venerabilis comes Rogerius et devota ejus conjux Adelaisia nomine nostra ætate Deo obtulerunt.

Erat in suburbis civitatis Scropesbery parva satis ecclesia in honore beatorum apostolorum Petri et Pauli constructa cujus situs ad collocanda servorum Dei habitacula congruus videbatur. Comes ergo fretus licentia et auctoritate regis Willielmi et Cantuariensis archiepiscopi Lanfranci necnon et Cestriensis episcopi Petri, ad cujus diocesim isdem pertinebat locus, monachos illuc pro sua suorumque salute Domino exoraturos aggregavit anno ab Incarnatione Domini millesimo octogesimo septimo.

Quibus ad victus subsidia ipse et veneranda ejus uxor possessiones subscriptas imperpetuum donaverunt. Imprimis scilicet vicum unum eidem ecclesiæ contiguum cum tribus molendinis cunctusq. redditibus ad eum pertinentibus, qui vicus Anglicè dicetur Biforieta quod Latinè significat ante portam. . . .

Hæc omnia venerabilis comes Rogerius cum filiis suis Hugone scilicet Roberto Rogero et Arnulpho concessit atque firmavit.

Hæc omnia ego Willielmus rex Anglorum concessi et sigillo meo corrobaravi.

We are moved by the example of the holy fathers to sign our hand to a script preserving the memory of what possessions in our time we have known to be granted to the holy church by the faithful—as well because of the practical benefit of such a record (since by their example posterity may be moved to do good) as because of the known perversity of malignants who with one accord band together for rending asunder ecclesiastical possessions. We, therefore, imitating the industry of the worthies [the monks], have had the care to transmit for the attention of posterity a writing of a certain work of piety which the Venerable Earl Roger and his devoted wife, named Adelaisa, have [now] offered to God in our time.

In the suburbs of the city of Shrewsbury was a small, poor, little church, built in honour of the Saints Peter and Paul, the site of which was seen to be suitable for providing a small habitation for the servants of God. The earl, therefore, relying on the licence and authority of King William, the Archbishop of Canterbury, Lanfranc, also of Peter, Bishop of Chester, to whose diocese the place belonged, has congregated the monks there to exhort God for the welfare of the souls of himself and his, in the year of the incarnation of our Lord one thousand and eighty-seven.

As an auxiliary support for them, he and his revered wife have given in perpetuity the under-written possessions.

First, one road contiguous to the said church, with three mills

II. SHREWSBURY ABBEY MILLS.

3. An alleged Forged Charter, 1094.

and the whole revenues appurtenant to them: this road in English is called Biforieta, which in Latin signifies Before the Gate. . . .

The whole of these the Venerable Earl Roger, with his sons, Hugo, Robert, Roger, and Arnulph, has conceded and confirmed.

All the foregoing I, William, King of the English, have conceded and confirmed by my seal.

Here we clearly see all the various gifts given from 1083 to 1087 duly entered as of 1087, the year when, on the appointment of abbot and chapter, the abbey, "by consent of King William," had officially come into existence, and its complete foundation charter had been formally handed to it by the founder. The specific grant of the city multure is not said to have been made, but generally "all known rights" appertaining to the mills. Doubtless this originally was the precise term used in what is said to be the "first" grant of mills in 1083; it being only the formal charter at the close of 1087 that had clearly stated what those "known" rights included in the eyes of Earl Roger—namely, the multure of the town.

The same version of the milling terms appears in Fundationis et Dotationis Historia, compiled by the monks about a century later:—

Monast. Ang., i. 376[a] (Ex. Reg. Abb)

Rogerus et Adelaisa concedente rege Willielmo et archiepiscopo Lanfranco et patriæ præsule Petro, posuerunt monachos Deo servituros in quadam ecclesia in honore sanctorum Petri et Pauli constructa in suburbiis Salopesbiriæ civitatis, qui domino exorarent assiduè pro animarum suorum et hæredum qui ante eos ipsum honorem rectè possederunt vel possessuri sunt.

Hanc autem ecclesiam, memoratus, comes et comitissa de suis proprius facultatibus à fundamentis instaurantes, habitationi monachorum congruum efficere inceperunt sed morte interveniente non perfecerunt. Huic vero ecclesiæ ea quæ infra continentur tam pro animarum suarum quam pro omnium Christianorum salute donaverunt et perpetuoque concesserunt.

Primo scilicet dederunt congregatis illuc fratribus vicum quendam juxta eandem ecclesiam positum cum tribus moléndinis cunctisque redditibus que ad eam pertinere noscuntur. Hic autem vicus dividitur à civitate Salopesbiriæ solummodo alveo fluminis intercurrentis cui nomen est Sabrina. Vicus autem ipse Biforietta vocatur quod nos lingua Gallica ante portam dicimus. . . .

Hæc autem omnia acta sunt anno ab incarnatione Domini mlxxxvij.

II. SHREWSBURY ABBEY MILLS.

3. An alleged Forged Charter, 1094.

Roger and Adelaisa, with the consent of King William, the Archbishop Lanfranc, and the principal father [of the diocese of Chester] Peter, placed the monks, servants of God, in a certain church, built in the suburbs of the city of Shrewsbury, in honour of the holy Peter and Paul, that assiduously they might exhort the Lord for the souls of them and their heirs; the said monks rightly possessing such honour, before [as being of Roger's monastery at Seez], even as they are its possessors now.

The earl and countess, who, be it remembered, out of their own resources were rebuilding this church from its foundations, commenced the erection of a suitable habitation for the monks; but death intervened, and they did not complete it.

To this church, for the welfare of the souls as well of themselves as of all Christians, they gave and conceded in perpetuity as herein contained.

First, they gave to the congregation of brethren there a certain road situated near the said church, with three mills, and the whole revenues known to appertain thereto. This road is separated from the city of Shrewsbury only by the course of the river, called the Sabrina, running between. The road is called Biforieta, which in our Gallic tongue we term Before the Gate.

The whole of these acts were of the year of the incarnation of our Lord 1087.

This second recital of the "first" grant again omits any mention of the city multure; and, as before, "the whole revenues appurtenant to the mills" are claimed—including the town multure by Roger's uncontested charter. No doubt, had the scribe who wrote the foregoing (or the abbot who drafted the charter of William II., from which the transcriber evidently made this copy) been able to foresee the trouble shortly to arise over what they (like their recent historians) may have deemed mere "petty, minor details," we should have had in each case a more precise statement of what the rights of these mills were.*

* Another mill (one of Saxon origin, and still existing) on the same Meole brook, as it is now called, in the suburbs of Shrewsbury, is that of Meole Brace. In Saxon times it gave name to the hamlet, Mela; this being held by Edith, queen of the Confessor, and the mill being scheduled in Domesday at the high valuation of 20s. per annum. During very early times the mill seems to have been a malt mill (mola brasium); hence the present name of the village, Meole Brace. Throughout all the contests between the abbey and the burgesses the name of this mill does not occur; because on one hand it was outside the city, and on the other it was beyond the lands of the abbey. Still, it certainly would be one of those to which the disaffected were wont to repair during the continuance of the conflict.

II. SHREWSBURY ABBEY MILLS.

4. City Multure re-granted, *c.* 1120.

4. The next evidence is a charter by Henry I.; and whatever opinion we may entertain as to the abbey possessing the multure of the city in virtue of Roger's charter, there can be no doubt as to its grant, or rather re-grant, by Henry in his own personal right as Earl of Shrewsbury. Owen and Blakeway state that this charter remains in the chartulary, but, as with most of the other of these documents which they discuss, they do not print it. They state that it is granted by "rex Henricus comitatum in proprio tenens domino," acting, as we have said, in his capacity of manorial lord; and conferring on the abbey the manorial milling rights of the town. The following is the clause bestowing these:—

Monast. Ang., i. 382 (Fund. et Dot. Hist.)

Omnia ista suprascripta gloriosus rex Henricus coram omni consilio suo apud Londoniam se proprio signo et sigillo confirmavit. . . . Concessit quoque monachis molturam totius civitatis Salopesbiriæ præcipiens ut nullus in toto circuitu civitatis nec in uno ponte nec in alio molendinum faceret nisi monachi vel cui ipsi concesserunt.

Testibus Richardo Londiniæ episcopo &c.

All the foregoing the illustrious King Henry [I.], before his entire council in London, with his signature and seal confirmed. He conceded also to the monks the multure of the whole city of Shrewsbury, directing that within the entire circuit of the city, neither at one bridge nor the other, none should make a mill except the monks, or any whom they have authorised.

Witnesses, Richard, Bishop of London,* &c.

5. Grant confirmed, 1227.

5. In 1227 Henry III. fully confirmed this grant of city multure by Henry I.:—

Monast. Ang., i. 382b (Ex. Reg. Abb.)

Henricus &c. Sciatis nos concessisse &c. abbati et monachis Salopesbiriensis ecclesia has subscriptas terras et decimas et ecclesias: . . .

De dono comitis Rogeri vicum unum juxta eandem abbatium scilicet Biforegatam cum omnibus molendinis et redditibus ad vicum illum pertinentibus. . . .

---

* One of Earl Roger's sons, mentioned elsewhere in the *Historia* as "Richardus de Belmesio qui postea Londinæ episcopus factus est."

II. SHREWSBURY ABBEY MILLS.

De dono regis Henrici avi H. regis avi nostri piscarium unam subtus pontem Salopesberiæ et totam molturam ejusdem civitatis: et quod nullus faciat molendinum in Saverna et territorio ipsius urbis nisi per monachos præfatos.

5. Grant confirmed, 1227.

Data per manum venerabilis patris Radulfi Cicestrensis episcopi cancellari nostri apud Westmonasterium xxv die Junii anno regni nostri undecimo.

Henry, &c. Know ye that we have conceded to the abbot and monks of the church of Shrewsbury the under-mentioned lands, titles, and churches: . . .

By gift of Earl Roger, one road beside the said abbey, that is the Before-the-Gate, with all mills and revenues appurtenant to the said road. . . .

By gift of King Henry, ancestor of King Henry, our ancestor, one fishery under the bridge of Shrewsbury and all the multure of the city: so that none may make a mill on the Severn or the territory of that town except the said monks.

Given by the hand of the Venerable Father Randulph, Bishop of Chichester, our Chancellor, at Westminster, June 25, in the eleventh year of our reign [1227].*

6. Monastic Soke broken, 1267.

6. At this early period Shrewsbury Abbey had the misfortune to be deprived, by the influence of Henry III., of the valuable milling rights which he himself had confirmed. It is only at this stage that the local historians, Owen and Blakeway, take up the story. Ignoring the earlier portion we have recounted, and reciting but a few of the facts of the sequel, they misquote these; publish nothing of the actual documentary evidences of the case; and exhibit a complete misapprehension of the entire question at issue. Like most other historians of their day, they give but little heed to the fortunes of early feudal corn milling; and the semi-apologetic prologema with which they commence their scant paragraph on the subject only emphasises the more the present desirability of endeavouring to exhibit this typical struggle

* "Charters are drafted by the abbot and monks, the grantees; and the document is then offered for royal consideration, perhaps accompanied by vouchers, but more certainly with a pecuniary recommendation." — Antiq. of Shrop., Eyton, i. 28.

"The practice of kings granting charters in the text of petitions ultimately proved unsatisfactory; being partly remedied by stat. 1 Henry IV., c. 6, and more fully by an enactment in the next Parliament; since which time all royal grants are issued 'ex speciali gratia, certa scientia, et mero motu regis,' of the especial grace, certain knowledge, and mere motion of the king." — Rotuli Chart., Turr. Lond. intro.

II. SHREWSBURY ABBEY MILLS.

6. Monastic Soke broken, 1267.

Hist. Shby., i. 129.

in its true light. Its commencement, we are told, "must be mentioned," as it is "a domestic transaction which throws some light on the state of society in Shrewsbury in the thirteenth century." But, peradventure, "in discussing it the local historian is obliged to descend from the dignity of his narrative, from princes and nobles and the infraction of national treaties, to the minor details of daily life and the adjustment of petty differences." Yet, it may be that the story is not without a certain dignity of its own, in its delineation of the monastic mill-owners striving during successive centuries to protect their perfectly legal chartered rights against spoliation by the populace; while, unfortunately for the abbey, there is certainly not lacking even an allusion to kings and the infraction of treaties.

Between 1227 and 1267 the burgesses of Shrewsbury, bound to grind at the manor mills, endeavoured to break the soke, and erected private mills. As the historians grandiloquently observe, the burgesses were "prompted to spurn the slavish badge of their former subjection"—a statement which, in any case, does not correctly describe a revolt against the law of the land and the rights of property. The abbot, not being manorial lord, had not himself the right to destroy the illicit mills, and accordingly brought an action at Salop Assizes to restrain their owners from using them. The hearing was pending, when, in the autumn of 1267 (Henry III. being then in Shrewsbury, intent on promoting the loyalty of the great border city), the case was withdrawn "by the action of common friends"; and a so-called "compromise" was made. By this the abbey absolutely lost its manorial milling rights; and there can be no question who the "common friends" were. The fact is that at the time William de Upton, abbot, was in disgrace at

II. SHREWSBURY ABBEY MILLS.

6. Monastic Soke broken, 1267.

Court; and Henry III., with the concurrence of various members of his council, seems to have retaliated upon the monks by peremptorily quashing their pending action, and permitting their spoliation in the interests of the burgesses, who were then in high favour with the sovereign. This punishment having been decided upon, Henry, on September 23, 1267, "graciously remitted to the said abbot and convent all the rancour and indignation which he had conceived against them by occasion of the trouble of the realm; and pardoned all the transgressions which they are said to have done against him by adhering to S. [Simon], quondam Earl of Leicester"; and at the same time he extracted from them for a purely formal favour £700 (present value). It is, thus, quite clear by whose influence it was that the monks had agreed in August to the one-sided so-called compromise. The historians say "the suit must have terminated in their favour, for the words of Henry's charter to them are express; but the discreet ecclesiastics probably saw that the spirit of the age was too far advanced to make it prudent for them to press in its full rigour a monopoly so hateful." It is patent that it was not the "spirit of the age" that was adverse to the monastic mills, but the spirit of the burgesses of Shrewsbury. The "spirit of the age" advocated the then beneficial scheme of the wealthy and powerful providing a sufficiency of properly equipped mills throughout the country. In virtue of that spirit the kingdom was filled with such feudally supported mills; and prince and squire, monk and secular, even Henry III. himself, carefully maintained them, and insisted upon their tenants using them. What the discreet ecclesiastics really did see when they agreed to give up their rights was the menacing figure of Henry III. in the background of their rebellious neighbours.

II. SHREWSBURY ABBEY MILLS.

6. Monastic Soke broken, 1267.

Trans. Shrop. Arch. Soc. 1894, 341.

The actual record of this affair and of its sequel is not quoted by the historians, and antiquaries are therefore much indebted to the Rev. C. H. Drinkwater, of Shrewsbury, for a transcript of the interesting document. The essential portions of this are appended; commencing with the terms of the "compromise" of 1267:—

Cum inter religiosos viros abbatem et conventum Sancti Petri Salop ex una parte et communitatem burgensium ville Salop ex altera motum esset placitum in curia Domini regis de quibusdam molendiniis per predictos burgenses in eadem villa et extra levatis contra cartas et libertates predictorum abbatis et conventus a diversis regibus et magnatibus Angliæ eisdem concessas secundum quod in cartis ipsis plenius continetur. Cum dicta burgenses tria molendina ad equos et unum ad ventum intra muros dicte ville et tria molendina ad aquas * in suburbio ejusdem villæ levassent:

Tam iidem burgenses quam predicti abbas et conventibus concesserunt quod predicta tria molendina in dicto suburbio levata prosternantur et in perpetuo prostrater remaneant. Et quod predicta quatuor molendina infra muros predictos existencia in locis in quibus nunc sunt in perpetuum remaneant vel ad loca competenciora in eadem villa de communi consensu partium predictarum sumptibus communibus transferantur levantur et sustententur. Ita quod ipsa quatuor molendina secundum statum in quo nunc sunt per duodecim probos et legales homines juratos neutri parti suspicatos fideliter apparcientur. Et quod dictus abbas et conventus medietatem dicti precii dictis burgensibus refundant si molendina ipsa ibidem debeant remanere: sic autem alibi de communi consensu de communibus sumptibus utriusque partis leventur et sustententur. Omnes vero exitus et proventus ipsorum quatuor molendinorum inter partes predictas equalibus porcionibus sine contencione dividantur; et ipsa molendina et molendinarii et alii ministri in ipsis molendinis communi concensu consortietur sumptibus communibus utriusque partis sustentatur.

Preterea de communi consilio et consensu tam predictorum abbatis et conventus quam predictorum burgensium provisum est et concessum quod duo molendina aquaria per aquam Sabryne currencia circa villam Salop unacum stagnis eorundem sumptibus communibus construantur et sustententur ubi utraque pars ad commune commodum magis viderit expedire. Ita quod omnes exitus et proventus ipsorum molendinorum sicut et aliorum quatuor antedictorum equaliter inter partes dividantur. . . . In omnibus autem predictis molendinis consortuantur molendinarii et alii ministri tot ex parte predictorum abbatis et conventus et eorum successorum quot ex

* The terms "equos" and "aquas" seem by the context to be relatively applied as we have given them, though they are not so in the transcript. The verdict of the jury also will be seen to mention the horse-mills as being in the town.

parte burgensium et eorum heredum. Ita quod de communi consensu utriusque partis sustententur vel amoveantur et alii loco ipsorum sustituantur.

6. Monastic Soke broken, 1267.

Si vera omnia molendina predicta unacum molendinis predictorum abbatis et conventus in suburbio constructis ad multuram dicte ville non sufficiant et plura molendina de communi concensu levare voluerunt, tunc tam de sumptibus quam emolumentis eorundem molendinarum per omnia fiat inter partes, sicut de aliis molendinis communibus ut predictum est.

Nullus vero de homnibus Salop impediet aliquem quominus ad communia molendina predicta et ad molendinum predictorum abbatis et conventus in suburbio et alibi pro voluntate sua propria libere possit ire et ad ea molere.

In cujus rei, &c., Domino Walter de Merton; Magistro Godefrid Gyffard, tunc cancellario Domini regis; Magistro John de Chyshull, Domino Robert Walerand, John de la Lynde, Robt. Anguyllon, et aliis magnatibus tunc apud Salop existentibus.

Whereas between the religious men the abbot and convent of St. Peter of Salop on the one part, and the commonalty of the burgesses of Salop on the other, motion is made in the court of the lord the king regarding certain mills erected by the said burgesses both within and without the said town contrary to the charters and liberties of the said abbot and convent by divers kings and magnates of England conceded to them, as fully shown by their charters; and whereas the burgesses have erected three horse-mills and one windmill within the walls of the town and three watermills in the suburb of the same:

The said abbot and burgesses have agreed that the three mills in the suburb shall be pulled down and shall perpetually remain prostrate. And that the four mills within the walls shall remain in perpetuity either in the places where they now are or in any other suitable places in the town, and by the common consent of the said parties shall be raised, maintained, or transferred at their joint expense. The said four mills shall be faithfully viewed, with regard to the condition in which they now are, by twelve honest and legally qualified men, not suspected of bias toward either party. If the mills remain where they are, the abbot and convent shall refund to the town one-half of their assessed value; if they are removed by the common consent of both parties, they shall be rebuilt and maintained at their common cost. All expenses and receipts of the said four mills shall be equally divided without contention between the parties; and the mills, with the millers and other servants therein, shall be arranged and maintained by the common consent and at the common cost of the parties.

Furthermore, it is provided and conceded by common counsel and consent of the abbot and burgesses that on the river Severn round about the town of Shrewsbury shall be erected and sustained at the expense of both parties two watermills with their dams at such places as may seem to both parties most convenient; and all the expenses and receipts of those mills, as of the other four already mentioned, shall be equally divided between the parties. The millers and all the servants in the aforesaid mills to be appointed as well on behalf of the abbot and convent and their successors as of the burgesses and their heirs, and the mills to be so sustained or removed to other sites by the common consent of both parties.

6. Monastic Soke broken, 1267.

If it happen that all the aforesaid [six] mills, together with the [three] mills of the abbot and convent in the suburb [of Abbey Foregate], should not be sufficient for the multure of the entire city, and more mills are by common consent desired, then the costs as well as the profits of such new mills shall in all things be dealt with between the parties as is already set forth regarding the other common mills.

None of the men of Shrewsbury shall hinder any one going to the said common mills or the mills of the abbot and convent in the suburb [of Abbey Foregate] or elsewhere at their own free will to grind.

Witnesses: Walter de Merton; Master Godfred Gyffard, then the chancellor of the lord the king; Master John de Chyshull, Lord Robert Walerand, John de la Lynde, Robert Anguyllon, and other great men then dwelling in Salop.

Briefly stated, the result was that the abbey was deprived of its sole multure rights, and was made a partner with the town for the maintenance of common mills. Theoretically this arrangement was for the public benefit, since it broke a monopoly; but practically it proved a public disadvantage; for the burgesses soon failed in the fundamental duty which the monkish manorial millers had necessarily, if but for their own sake, carefully observed—that of maintaining the mills in an efficient state. Thus closed the events of 1267.

7. The Mill on the Gaye, 1268.

7. In 1268 the burgesses are again found in conflict with the abbey. The garden of the monastery lay upon the bank of the Severn, and comprised an old meadow called the "Gaye" and an island in the stream closely adjoining it. Here the monks in this year built a mill for the private use of their house and of their own little mesne manor of Abbey Foregate. The mill lay below the town on the course of the river, and was passed by all boats trading up from the Bristol district. Hence its existence threatened

II. SHREWSBURY ABBEY MILLS.

7. The Mill on the Gaye, 1268.

the safety of navigation; and, in accordance with the universal ancient laws and usages dating from Roman days, the burgesses exercised the right of pulling it down. The abbot thereupon agreed to avoid causing any hindrance in future, and an amicable agreement was made for its restoration. Its erection had been no breach of the arrangement made in 1267, nor did the burgesses ever make any complaint of its being such. Yet what say the historians? "In 1267 the abbey had agreed to erect no more new mills otherwise than conjointly with the corporation; yet within a year they did erect one on their land, the Gaye; . . . and the burgesses destroyed it, as a plain infraction of the agreement":—an assertion quite disproved by the record they themselves quote:—

Text, I. 41, II. 182.

Hist. Shby., ii. 99.

Notum sit omnibus ad quos pressens scriptum pervenerit quod omnes lites tunc mote et omnes controversi inter dom. abbatem et conv. Salop et burgenses burgi Salop ratione meremi destructi molendini aquatici faciendi in terra dict. abb. et conv. que dicetur le Gaye super Sabrinam et occasione alia quacunque, conquieverunt in hunc modum: viz. quod dicti burgenses communiter unanimiter et fideliter promiserunt quod nullum impedimentum nocumentum impenetrationem vel occasionem facient, quin dicti abbas et conv. in loco predicto sc. le Gaye molendinum aquaticum ad libitum suum et aisiamentum possint facere, et eo facto ad commodum suum libere uti. Si autem aliquod impedimentum evidens ratione dicti molendini vel alicujus ob id. f'c'i. vel appositi vel al aliud pertinentis quod rationabili modo et rationabiliter potest ostendi fuerit factum ad nocumentum dicti burgi, ob quod naves majores et minores libere ad dictum burgum venire et redire non possint, dicti burgenses omnes terras redditus et tenementa dictorum abbatis et conv. ubicunq. fuerint inhabitate burgens in manu sua sine contradictione et cavilatione tenebunt: Et interim omnes fructus et proventus ex dictis terris red. et ten. provenientes plenarie suscipient donec emenda talis transgressionis facta fuerit. In quorum test Duo. Thoma Corbet, Roberto de Gyros, Will° de Hedley. Dat die S. Luce Evang. A. gr'e mcclx octavo.

Ibid., ii. 100.

Be it known to all to whom these presents shall come: That all contest now raised and all controversies between the lord abbot and convent of Shrewsbury and the burgesses of Shrewsbury, regarding the destruction of the timbers of the watermill built on land of the said abbot and convent, called "The Gaye," on the Severn, and all other disputes, have been settled in this manner: viz. the burgesses

II. SHREWSBURY ABBEY MILLS.

7. The Mill on the Gaye, 1268.

have conjointly, unanimously, and faithfully promised to make no injurious impediment to the undertaking; so that the abbot and convent may at pleasure make at the aforesaid place, viz. The Gaye, a watermill and easement, and, that being done, may freely use the same at his convenience. If, nevertheless, any impediment —evidently by this mill, or any other by the said abbot made or placed, or by other things, being appurtenances—may reasonably be apparent to be done [on the river] to the injury of the town, so that vessels, large and small, cannot come and return freely to the said town, then the burgesses shall take into their own hands all lands, rents, and tenements of the said abbot and convent, wherever they shall be inhabited by burgesses, and shall hold the same without cavil or contradiction, and shall fully receive all issues and profits arising from the same till satisfaction for such transgression be made. Witnesses, the Lord Thomas Corbet, Robert de Gyros, and Wm. de Hedley: St. Luke the Evangelist's Day [October 18], in the year of grace 1268.*

8. Compact neglected by the Town, 1280.

8. We may now see how the enforced milling partnership between the abbey and the town progressed; though the local chroniclers, having incorrectly referred to it thus far, make no further allusion to it. The burgesses, with none of the precision and officialism of the monks, contrived in but a few years to get their share of the business into complete disorder. The four common mills in the town they had refused to join in repairing; even after the abbot, in accordance with law, had formally called upon them to do so. In due course the mills became ruinous, and to make up for their loss some of the burgesses erected a dozen others. These necessarily were flimsy, inconsiderable structures, mostly horse- or ass-mills; as a dozen mills of any size were obviously far beyond the requirements of the town. Such a state of affairs no doubt abolished all exclusive soke for any one, but it also abolished all compulsory liability upon any one to maintain, year in and year out, for the common benefit, adequate and permanent mills grinding at a legally stipulated toll—a vital responsibility which feudal soke law had wisely been intended to secure,

Text, III. 74.

* "The site of this mill is still [1825] traditionally remembered. A pavement which had belonged to it was taken up a few years ago."—*Owen and Blakeway.*

II. SHREWSBURY ABBEY MILLS.

and for which as yet no sufficient substitute was forthcoming. At this juncture, to save his share in the property from ruin, the abbot, about 1279, appealed to Edward I.; who issued a writ, stating "he had been informed on the part of the abbot that the burgesses did not preserve the compact, and had permitted the four mills in the town to fall in ruin for lack of maintenance due on their part, while they had also set up in the town, without the consent of the abbot, twelve other mills."* He ordered all the parties to attend in the court of the king wherever he might be in the fortnight of the festival of John the Baptist (June 24). It was Hilary Term, 8 Edward I. (1280), before the case came before a jury, and a verdict was then returned on all the counts in favour of the abbey.

8. Compact neglected by the Town, 1280.

Jurati dicunt super sacramentum suum quod tria molendina ad equos et unum molendinum ventriticum sunt diruta et prostrata ob defectum construcionis et reparacionis predictorum burgensium communitatis predicte et non per defectum ipsorum abbatis et conventus. Et de quarto molendino dicunt quod omnino ignorant ubi situm fuit nec de eo unquam noticiam habuerunt. Et dicunt quod tria predictorum molendinorum nunc sunt edificata per predictos burgenses: et inde proficium capiunt. Dicunt etiam quod

Trans. Shrop. Arch. Soc., ut supra.

* In July 1403, on the day after the battle of Shrewsbury, Henry IV., erecting a mock pillory of stones from these mills in the market-place, exposed and quartered there the body of Hotspur:—

Henricus Percy non solum semel occidit sed quantum in ipso est bis et ter interfecit. Quia postquam semel fuit occisus et domino de Fernyvale ad sepelendium traditus et liberatus, qui ipsum ecclesiasticæ sepulturæ prout moris erat christianorum cum honore quo tunc potuit tradidit, et cum suffragiis mortuorum missarum et aliarum orationum ipsius animam apud Deum commendavit; idem dominus Henricus, ut cruentia bestia, ejus sanguinem denuo sitiens, ejus corpus de tumulo exhumari et extrali præcipit; et inter duos molas asinarias in quodam vico de Sciropesberie juxta collistrigium reponi et sedere fecit, ac cum armatis hominibus custodiari; postmodum decollari et membratim dividi et quarterisari, et caput et ejus quarterias ad regni certas civitates transmitti jussit.

Anglia Sacra. artic. R. Scrope 366; Archeologia, xx. 162.

Henry slew Percy not once for all, but—so far as in him lay—annihilated him twice or thrice. For after he was slain and his body delivered and released for burial to Lord Furnival, and after he, with such honour as he then could, had given it sepulture with the rites of the Church, as the manner of Christians was; and, with the last offices for the dead, masses, and other prayers, had commended his soul to God; the said Lord Henry, with the cruelty of a beast, thirsting for his blood anew, ordered the body to be exhumed and taken forth. Between two ass-mill stones beside the pillory in a certain street in Shrewsbury it was placed and made to sit, guarded by armed men; afterwards it was beheaded, its limbs severed, and quartered; the head and quarters being ordered to be transmitted to certain towns in the kingdom.

II. SHREWSBURY ABBEY MILLS.

8. Compact neglected by the Town, 1280.

predicte burgenses levaverunt duodecim molendina in predicta villa Salop contra formam composicionis predicte ad dampnum ipsius abbatis trescentarum librarum.

Questio si predictis abbas unquam eosdem burgensos monuerit ad reparacionem et sustentacionem predictorum molendinorum faciendorum, dicunt quod sepius eos monuerit, quoniam hos omnino facere recusarunt. Questio si predicti burgenses unquam prefatos abbatem et conventum monuerunt ad reparacionem et sustentacionem predictorum molendinorum pro porcione ipsis contingente, dicunt quod non. Questio cujusmodi molendina sunt illa xij molendina que prefati burgenses levaverunt, dicunt quod molendina cum rotis et similibus aliis instrumentis ad molendinum pertinentibus et que molere possunt viginti quarteria bladi nocte et die ut intelligunt.

Ideo consideratum est quod predicta tria molendina reparentur et releventur sumptibus predictorum burgensium: et cum reparata et relevata fuerint sumptibus utrorumque abbatis et conventus et predictorum burgensium sustententur juxta formam composicionis predicte: et quod predicta duodecim molendina per prefatos burgenses levata sumptibus predictorum burgensium per vicecomitem prosternantur et prostrata permaneant. Et quo ad dampnorum, &c., dies datus, &c., in tres septimanas, &c., quo ad audiendum judicium, &c.

The jury find upon oath that three horse-mills and one windmill [in the town] are broken and prostrate by reason of the default of the commonalty regarding their maintenance and repair, and not by default of the abbot and convent. Of the fourth mill they say they are entirely ignorant where its site may have been, and they have had no evidence on the point. Three of the aforesaid mills are now rehabilitated by the burgesses, and they take the profits therefrom. Also the burgesses have erected twelve mills in the town of Shrewsbury contrary to the terms of the compact and to the damage of the abbot and convent to the amount of £300.

Questioned if the abbot had ever admonished the burgesses to attend to the maintenance and repair of the mills, the jury say that he frequently did so, but they all refused to do it. Questioned if the burgesses had ever admonished the abbot and convent respecting their share of the maintenance and repair, they find that they did not. Questioned what manner of mills are those twelve which the burgesses have erected, the jury say that they are mills with wheels and similar other instruments appropriate to mills; and that, night and day, they could grind twenty quarters of grain, as they understood.

Therefore the decision [of the Court] is that the three mills should be repaired and restored at the cost of the burgesses, and when repaired and restored should be maintained at the common cost of the burgesses and the abbot, according to the tenor of the aforesaid compact; also that the twelve mills erected by burgesses should be pulled down by the sheriff at their cost, and should permanently remain prostrate. The question of damages as may be decided at a further hearing by the Court in three weeks' time.

II. SHREWSBURY ABBEY MILLS.

9. Town Soke abolished, 1326.

9. The foregoing verdict, in will be remembered, was delivered in January 1280. The Exchequer record from which it is quoted affords evidence that even then the conflict was not ended; for the whole story, as it is here analysed, is recorded, not in documents of 1267 nor of 1280, but in one drawn up in 1307, entitled "Placita coram Domino rege apud Westmonasterium de termino Sancte Trinitatis anno regis Edwardi filii regis Henrici trecessimo quinto"; these being the pleas for a trial before Edward I. in 1307, when those of the earlier suits, as given above, were recited. What the result of this action was does not appear, though from subsequent events we may reasonably conjecture that the burgesses were again found in default, and ordered to keep to the terms of their partnership with the monks. For in 1326 they endeavoured to get rid of the partnership altogether, and oust the monks from their last share in the multure of the town. "Encouraged by an extraordinary mark of royal notice, and by a confirmation of their old charters [in 1323], the corporation ventured shortly after, namely, at the Parliament holden in the nineteenth year of Edward the Second's reign, to remind him of his promise of a new charter, and to solicit further indulgences." Among these latter they boldly asked permission to make "engyns de moudre lour bleez et bretz"—engines to grind their corn and malt. But in the meantime the barometer of the royal favour had gone down again, and Edward II. did not accede to their requests. In the following January Edward III. came to the throne, and a favourable occasion ere long presenting itself, the citizens petitioned the young king and obtained at last, on May 12, 1328, the sovereign's licence to erect within the town the desired "ingenia per que blada et brasia molere possent." Thenceforward any one in

Hist. Shby., ii. 158.

II. SHREWSBURY ABBEY MILLS.

9. Town Soke abolished, 1326.

the city could erect a horse-mill or other mill for his own use or for hire, or could use a hand-mill at his own private house: all compulsory soke in the town, whether in favour of burgesses or monks, being abolished.

Still the matter does not seem to have ended. Evidently the abbey made another effort to secure its old rights; for it appears that the above documents of 1267, 1280, and 1307 are additionally

Mill and Ruin of Monastery, Shrewsbury.

endorsed Factum per recorda anno nono H. Quinto coram eo:—Done by record 9 Henry V. before him. In the year 1422, therefore—a century and a quarter after the litigation of the time of Edward I.—the old contention once again had been raised, and all the pleas of the former trials once more had been produced at the court of the king by the abbey, in support of what little remnant of milling right remained to it; namely, the right of grinding for its own suburban district.

II. SHREWSBURY ABBEY MILLS.

10. Reliquiæ.

10. Thus after the lapse of three and a quarter centuries, largely spent by the abbey in endeavouring to utilise its intangible charter rights over the town, we find its power reduced again to the limits of the original provisional grant by Earl Roger—the milling of its little rural estate of Abbey Foregate. This it

The Mill-race, Abbey Foregate, Shrewsbury.

seems to have more or less retained till the Dissolution; when, the mesne manor of Abbey Foregate being acquired in 1558 by new owners, the three mills, dating from Domesday, were kept going till even their fragmentary soke over the countryside expired. Their ruins were to be seen, within the last score years, not far from the abbey church; one of them closely adjoining the infirmary of the monastery, and the others farther up the still existing race on the south side of the oft-mentioned road, Abbey Foregate.

DEE MILLS AND CAUSEWAY, CHESTER.

CHAPTER III.

DEE MILLS, CHESTER.

III. DEE MILLS, CHESTER.

1. Forewords.

1. THOUGH corn mills have stood on the bridge of Dee, Chester, at least eight centuries, their origin is not to be traced back to still earlier times, as may certainly be done in the case of very many of the smaller milling establishments of the kingdom. The bridge itself dates back very clearly to Saxon days. True, Ormerod states "the earliest mention of a bridge at Chester occurs in the MS. chronicle of St. Werburgh, which mentions its total downfall in 1227." Yet we may read in Domesday that a hide and a half of city land lay "beyond the bridge"; and that, according to Saxon custom, the citizens were bound to find labour for the repair of "the city wall and the bridge." But no allusion is made to any mills being on the bridge in Saxon times, nor till the year 1093.

Hist. Cheshire, 1882, i. 362.

Domesday, i. 262[b], 263[b]

These mills, of undoubted antiquity and considerable magnitude, and at one time, so far as the records go, the most valuable king's mills in the realm, are intrinsically worthy of a record of some approach to reliability; and it is satisfactory, therefore, that the effort to put some such record together has revealed an amplitude of material in the ancient records of the city of Chester, and in the mass of documents relating thereto comprised in the Harleian MSS. at the British Museum.*

* The extensive and interesting archives of Chester have been fully consulted on behalf of this sketch, by courtesy of the Corporation, aided by the good offices of Mr. S. Smith, Town Clerk. The appreciative care with which the valuable and unique records of the ancient city are preserved in a specially built strong-room at the Town Hall demands very cordial recognition.

III. DEE MILLS, CHESTER.

1. Forewords.

Not a little misapprehension has prevailed in modern times regarding the origin of these mills and their connection with Chester Abbey. A Cheshire antiquarian publication, which happens to aptly summarise these perpetuated errors, may be cited :—

Cheshire Sheaf, May 1883.

Hugh Lupus granted to the monks a great number of privileges. Of these princely gifts, not the least in importance and value was the grant of Dee Mills and the fishery there. These and similar donations continued a source of revenue to St. Werburgh's all through the medieval period, and down indeed to the Reformation. Then Henry VIII., with the reckless benevolence of one who gives to others what is not his own, transferred the Dee Mills and other large properties to the newly erected cathedral of St. Werburgh. Early in the reign of Edward VI. the mills had been lost in litigation with Sir Robert Cotton. It was at this juncture that the chapter granted to Edward Plankney a lease, which has more or less affected the material interests of the old city for more than three centuries.

The fact is that Lupus gave the monks the site for a mill; but this, so far from continuing a valuable source of revenue to them down to the Reformation, soon passed out of their hands, and the mills early became "king's mills." Evidence of Henry VIII. granting them to the new cathedral is lacking: he did grant them to R. Brooke, however. Sir Robert Cotton gained the mills, not by litigation with the monks, but direct by grant from Edward VI. Finally, the dean and chapter did not lease the mills to E. Plankney, but only the tithes of the same; and this lease in no material way affected the interests of the city for centuries. This last misapprehension seems identical with a theory broached elsewhere :— "It seems probable that the gift of Lupus to the abbey had something to do with a claim raised by subsequent owners of the mills to the right of grinding for the citizens"; but the abbey during the very few years that it had a mill on the bridge did not enjoy the right of grinding for the city. In a word, the monks had practically nothing whatever to do with Dee Mills during the entire period of their medieval history.

Text, post, p. 86.

Hist. Soc. of Ches., i. 238.

III. DEE MILLS, CHESTER.

2. It was in 1093 that the abbey mill was authorised to be erected on Dee Bridge. In that year Hugh Lupus, Earl of Chester, having founded the abbey of St. Werburgh in the city, joined with his countess in granting to it the usual gift of a mill:— 2. The Abbey and the Mills, 1093.

Dederunt huic ecclesiæ in ipsa civitate de suo dominico vicum a porta de North usque ad ecclesiam: et locum unius molendini ad pontem civitatis. Ormerod, i. 13.

They have given to this church in the city of the earl's own lordship the road from the North Gate to the church, and a place for one mill at the city bridge.

To obtain a head of water in the shallow Dee for the use of mills, Earl Lupus founded the stone causeway across the river, which in later centuries gave to the holders of the place very considerable trouble and anxiety. "The river of Dee was drawn unto the saide cittie with great charge by the said earle, or some of his predecessors before the Conqueste, from the auntiente course which it held before, a myle or two distant from the cittie; and a passage for it cutt out of a rock under the walls of the saide cittie; and the said earle also built the corn mills of Chester and erected the causey." There seems to be no early evidence that Lupus thus founded mills on the bridge besides the one which he empowered the monks to erect and work. Endowed with their new right, the monks built a mill for the use of their house, and exercised their multure rights over the residents on their land in Northgate Street. It was an awkward arrangement to compel tenants at North Gate to traverse the entire extent of the city with their corn to a mill outside the South Gate; and not one which consulted their convenience in the manner in which such arrangements were bound to do. Hence it is not surprising that it did not long endure. Harl. MSS., 2084. 157. Text, III. 218.

On the death of Hugh Lupus, his son Richard, the next Earl of Chester, shortly before 1119,

III. DEE MILLS, CHESTER.

re-granted his father's gifts to the abbey, with a notable addition :—

2. The Abbey and the Mills, 1093.

Ormerod, i. 17, 190.

Ego comes Ricardus post obitum patris mei dedi pro saluta animæ mea et suæ terram quæ fuit Wulfrici præpositi foris portam de North, prius per unam spicam frumenti deinde per unum cultellum super altare Sanctæ Werburgæ: et molendinum de Bache: et tres mansuras quietas et ab omni re liberas, duas in civitate et unam extra portam de North. . . .

Item Ricardus comes dedit decimum salmonem de ponte et locum unius molendini citra pontem et decimum illius molendini ultra pontem.

I, Earl Richard, after the death of my father, have given, for the welfare of my soul and of his, the land which was Wulfric's over against the North Gate, by service of first an ear of corn and afterwards of a knife or sickle upon the altar of St. Werburgh's; also the mill of Bache; also three messuages free from all dues, two being in the city and one outside the North Gate. . . .

Item: the Earl Richard has given the tithes of the salmon fishery at the bridge and of the place of one mill on this side of the bridge; also the tithes of that mill beyond the bridge.

The main consideration here is that, with extra land at the North Gate, Richard gave the monks the watermill of Bache in the same neighbourhood; and to that convenient locality there can be no doubt they speedily transferred the soke of their tenants from the bridge mill. He also granted, it seems to be stated, the tithes of the fishery and of the corn mill on the town side of the bridge, and of the mill on the other side of the same, this subsequently and probably then being a fulling mill. A corollary of this rearrangement of affairs appears incidentally in one of the later pleadings: "It doth appear by several records that in anno 1119 Richard, then Earl of Cheshire, granted a rent of 5[li] per an. to the then abbot of Chester for and in respect of the tithes of the said [bridge] mills, which rent is now [*c.* 1650] received by the Dean and Chapter." But if the abbey had owned the corn mill in 1119, it would ordinarily also have owned the tithe of the same. This grant of the tithe, therefore, may be taken as evidence that the monks had no

Harl. MSS., 2082. 6.

2. The Abbey and the Mills, 1093.

longer any proprietary interest in the mills; their single one apparently having been leased to or exchanged with the earl for this grant of its tithe and those of the adjacent properties at the bridge. This conjecture may be confirmed by another reference. In 1293 we find mention of a coroner's inquest touching the death of one Richard Molendinarius, who was "killed with a cogg wheele of a milne of the abbots of Chester, which milne was leased to the king." From the designation of the miller, it appears that he was a corn miller; fulling millers or walkers being rarely, if ever, so termed. The mill, therefore, seems to have been that originally granted by Earl Hugh, and subsequently nominally "leased" or granted back to Earl Richard by the abbey in consideration of the tithe; such lease or grant passing eventually to the Crown. However this may be, the mills in future appear as the property first of the earls and afterwards of the sovereigns, not of the monks; whose milling estate for themselves and their North Gate tenants lay at the North Gate, and comprised the watermill of Bache and a windmill they erected in later years in near proximity to the said gate.*

Harl. MSS., 2057. 8.

3. Claims by Lessee, 1278–1289.

3. Under the Earls of Chester the mills of Dee were constituted the manorial mills of the city (except the Northgate Street), and those Customs were established which in 1356 Edward III. confirmed, and which in later centuries were invariably recited in legal pleadings. Randolph Blundeville, sixth earl (1181–1232), seems specially to have placed the mills and their toll

* At the Dissolution the "survey of the said late Monastery" included:—

ij mills w'oute the Northgate of Cestr' w'in the pysshe of Saynte Oswald . . . iiij^{li} xiij^{s} iiij^{d}.

Subsequently these mills were worked by Alderman Bavand, and in 1567 an action was brought against his widow by the lessees of Dee Mills for infringement of soke rights. Bache watermill existed till modern times. The windmill was destroyed in July 1643; when, "to prevent the enemy making lodgement to the annoyance of the city, the great windmill without the North Gate was taken down."

III. DEE MILLS, CHESTER.

3. Claims by Lessee, 1278–1289.

and other rights on a sound and satisfactory basis; so much so that at the opening of the fifteenth century, at a period of popular dissatisfaction with the then management of the place, the customs of the time of "the late Earl Randolphe" were specifically quoted as good and laudable usages which should prevail for the benefit of the city.

On the death of the seventh Norman earl, John Scott, in 1237, Edward, the first Prince of Wales, became the next Earl of Chester. Under his rule as sovereign Dee Mills—now "king's mills"—speedily attained a very extensive and prosperous condition. Ormerod, the historian of Cheshire, quotes as the first of the keepers Richard Ingeniator,* 1302–1306. His name occurs somewhat before this date, however. An early document relating solely to the mills is a writ of Edward I., January 23, 1279, ordering the dismissal of Richard the Engineer from the mills. This lessee, like many of his successors, was a person of considerable local consequence; and the order for his removal—which was not carried out—seems not to have been due to any inefficiency or neglect on his part, but solely that Edward might grant the mills or their custody to the distant abbey of Vale Royal. A reason for this somewhat strange proceeding is to be found in the fact that Vale Royal, founded by Edward a few years before, when Prince of Wales, was a favourite House with him. Among other of his grants to it, made on attaining the throne, there may be found in the chartulary of Vale Royal one of land in Chester near the East Gate, and another of land in Bridge Street, the street leading direct to the mills; while, from the subsequent pleas of Richard the Engineer, it seems he had also granted away other

Hist. Cheshire, i. 225.

* A curious but not rare designation:—Terra Waldini Ingeniatoris (*Domesday*, i. 365[b]), Magistro Alberto Ingeniatori (*Cart. Ram.*, cxliv.). It had the general signification of a manager of works.

III. DEE MILLS, CHESTER.

lands in the precincts of the castle, these again being in close proximity to the mills. The writ under notice suggests that he contemplated granting the mills themselves to Vale Royal:—

3. Claims by Lessee, 1278–1289.

Harl. MSS., 2064. 256.

Edw: &c. Guncelyno de Baddelesm're Justic. suo Cestr. saltm. Quia nobis datur intelligi qd Ric' Lengyneer firmarius molendinoꝝ nroꝝ Cestr. firmā nō soluit ut deberet vobis mandams qd p'dcā molendina in manus nraš capiatis et alicuj ꝑbo et fidelī firmario ꝑ visum Leonis filij Leonis cam'arij nrī tradatis ad firmā; qui eidē cam'ar' nrō ad ops nrm̄ de vall Regali ꝑficiend fidelit' respondeat de firmā sup'dcā: Et p'fat' Ric' ꝑ arrer' firmā p'dcā p'fato camer' nrō solvend de die in die' distringatʒ. T. Meipsē apd Westm' 23 die Jañ anno rḡo nrī 7o.

Edward, &c., to Joceylin de Baddlesmere, his justiciary at Chester, greeting.* As information has been given us that Richard the Engineer, farmer of our mills at Chester, does not conduct them as he should, we send you mandate to take the said mills into our hands and deliver them to some other farmer, a person of probity and good faith, to farm under the supervision of Leo, son of Leo, our chamberlain; who, to our said chamberlain on our behalf for the profit of Vale Royal, shall faithfully answer for the said farm. And the said Richard for payment of his arrears to our said chamberlain shall be pressed from day to day. Witnessed by myself at Westminster, January 23, in the seventh year of our reign [1279].

Richard was not, however, ousted from his holding. As appears from another document, to be quoted, he held a twelve years' lease of the mills (with the fishery under the bridge) from the king, at a rental of £200, contemporary currency—an amount which, contrasted with any other mill-rents of those times, evidences the vast value to which the mills of Dee had already been brought. His lease had begun in 1277, as we know by his claim on its expiry in 1289; but before this latter date he had another claim to make, on account of the tenants of the Vale

* "Within this county palatine and the county and city of Chester there is and anciently hath been a principal officer, called the Chamberlain of Chester, who time out of mind hath had the jurisdiction of a Chancellor. The Court of Exchequer at Chester is the Chancery Court for the said county palatine, and the Chamberlain of Chester is a Judge in Equity. He is also a Judge in matters of common law. — Coke, 4 Instit., ch. 37.

"There is within the same city and county a Justice called the Justice of Chester, who hath jurisdiction to hear and determine matters of the Crown and of common pleas, as well within the county palatine as of the city of Chester." — Ibid.

III. DEE MILLS, CHESTER.

3. Claims by Lessee, 1278–1289.

Royal lands at Chester Castle refusing to grind at his mills. No contesting of his right under his lease of 1277 being possible, Richard claimed damages from the king who had introduced this element of discord into his soke district; and the sovereign, recognising his liability, ordered an enquiry to be held to assess the damages:—

Harl. MSS., 2083. 513.

Breve R. Edwardi I. Reginaldo de Grey Justiciario suo Cestriæ destinatum quo quia Ricardus le Enginur sibi suggesserat quod magna pars multure quam percipere debet de bladis in Dominicis Terris suis Cestriæ crescentibus, jam subtrahitur eidem, per assignationem quam fecit quibusdam hominibus partium illarum de Dominicis predictis, in escambium et recompensationem quarundam aliarum terrarum Abbati et Conventui de Valle Regali, in augmentationem sustentationis sui, per sese assignatum; volens inde certiorari ille mandat quod inquirat quantum dampni idem Ricardus sustinuerit annuatim per subtractionem multure predicte: et a quo tempore sic fuerit per dictam assignationem subtracta: et quantum multura illa sic subtracta valere potuit singulis annis a tempore subtractionis predicte: et quantum valere poterit ex tunc per annum: et qualiter et quomodo.

Writ of the King Edward I. sent to Reginald de Grey, his justiciary at Chester: for the reason that Richard the Engineer had preferred to him that great part of the multure he should receive on grain grown on the manorial lands of Chester was then withdrawn because of the assignment the king made to certain men of those parts of the said manorial lands in exchange and recompense for certain other lands assigned by him to the abbot and convent of Vale Royal, for their sustenance: the king therefore ordering enquiry to be made what damage Richard may have yearly sustained by withdrawal of the said multure; at what time in consequence of the said assignment it was thus withdrawn; what the withdrawn multure may have been worth each year since it was withdrawn; what it may be worth per annum for the future; and how and in what manner, &c.

Accordingly the justiciary held his court, and the verdict of the jury was forwarded to the king:—

Ibid., 2083. 514.

Inquisitio facta eodem anno: ubi per juratores compertum fuit quod a tempore quo Dns. Rex assignavit Dominicas Terras suas Castri Cestriæ hominibus partium illarum in recompensationem aliarum terrarum Abbati et Conventui de Valli Regali assignatarum, Ricardus L'Enginour per subtractionem multure quam de bladis Dominicis hominum predictorum percipere debuisset, sustinuit dampnum ad valentiam 58^{s} 7^{d} ob. Verum quod non potest eis constare de multura temporibus futuris.

3. Claims by Lessee, 1278–1289.

Inquisition held the same year: when it has been computed by the jurors that for the time the lord the king has assigned his manorial lands of Chester Castle to the men of those parts in recompense for other lands assigned to the abbot and convent of Vale Royal, Richard the Engineer by loss of multure on grain of manorial tenants aforesaid which he should have received has sustained damage to the amount of 58s. 7½d. He is not able to assess future multure to them, however.

The foregoing did not quite conclude the troubles of the lessee of the soke. In 1289 a William Fox purchased bread at Warwick and brought it for sale in a cart to Chester, where he was promptly arrested at the instance of Richard and conveyed to the castle, an action being brought against him by Fox; but on it being pleaded that the latter was acting contrary to the soke rights of the mills, the proceedings were quashed.

In the same year, 1289, Richard's twelve years' lease expired, and he duly claimed compensation from the king for damages sustained by several floods during his term:—

Harl. MSS., 2083. 513.

Breve R. Edwardi I. quo cum nuper commiserit Magistro Ricardo Ingeniatori suo, molendino sua Cestriæ et piscarum suam in aqua de Dee pertinentem ad pontem ultra aquam illam habend. usque ad finem 12 annorum: reddendo inde ad Scaccarium suum Cestriæ annuatim Ducentas Libras: et idem Ricardus, occasione inundationum aquæ predicta frequenter supervenientium, jacturam et dampna gravia sustinuerit, sicut dicit, mandat Reginaldo de Grey Justiciario suo Cestriæ, quod diligenter inquirat quantum dampni et jacturam dictus Ricardus passus fuerit occasione predicta: et si idem Ricardus inde culpa fuerit, nec ne. Teste &c. 7 die Septembris anno Regni 17.

Writ of Edward I. The king, having granted to Master Richard, his engineer, his mill of Chester with his fishing appurtenant, at the bridge crossing the Dee, for a term of twelve years, rendering therefor to the king's Exchequer at Chester £200 per annum; and the said Richard, because of frequently occurring inundations of the said river, having sustained breakages and serious damage at the mills, as he says; the king issued his writ to Reginald de Grey, his justiciary at Chester, directing that he diligently enquire what amount of loss and damage may have been inflicted on the said Richard, and whether or not he may have been to blame in the matter. Witnessed, &c., September 7, the seventeenth year of our reign [1289].

3. Claims by Lessee, 1278–1289.

An inquest was accordingly held, and the jury found in Richard's favour on all the counts, awarding him as total damages a sum of £154, for which the king was legally responsible, and which in due course was no doubt paid :—

Harl. MSS., 2083. 512.

Inquisitio facta eodem anno: quando juratores dixerunt quod postquam Dns. Rex. molendina sua et piscariam pontis Cestriæ Magro Ricardo Ingeniatori ad firmam 12 annorum tradidissit: primo anno ejusdem firme, tante inundationes aquarum frequenter supervenientium extiterunt, dum idem Ricardus stetit in servitio Dni. Regis apud Karnarvan, quod crates piscarie subtus pontem atachiari non potuerunt: et quod gurgites piscarie predicte et granaria cum parietibus molendinorum extirpati et asportata fuerunt: unde idem Ricardus dampnum et jacturam 40 librarum sustinuit. Item quod per tres annos sequentes dum idem Ricardus, per vices, in servitio Dni. Regis stetit in Wallia, per super-habundantes inundationes aquarum, gurgites et crates predicte, una cum calceto ibidem, dissipata et asportata fuerunt: unde idem Ricardus, quoad piscarium predictam, dampnum et jacturam ad valentiam 60 librarum incurrit. Item, quod per tempus quo idem Ricardus stetit in servitio Dni. Regis apud Trosselan, calcetum predictum per hujusmodi inundationes, dirutum et asportatum extitit; per quod molendina predicta fere per unum quarterium anni molare non potuerunt: unde per hoc et tales inundationes frequenter per totum tempus predictum supervenientes molendina predicta sepius absque molitione steterunt: per quod ipse Ricardus, quoad multuram per tempus predictum dampnum et jacturam ad valentiam 54 librarum optinuit.

An inquisition was held the same year, when the jurors found :— After the lord the king had granted his mill and fishery at Chester Bridge to Master Richard the Engineer to farm for twelve years, during the first year, when Richard was in the service of the king at Carnarvon, floods so frequently rose above the ordinary that it was impossible to fix the fish-weirs under the bridge, and the sluices of the fishery, the granary, with the walls of the mills, were extirpated and carried away, by which the said Richard incurred in losses and breakages £40. Also in three succeeding years, when Richard was several times in Wales in the service of the lord the king, various inundations scattered and carried away the sluices and weirs aforesaid, as well as the walking (or fulling) mill there; by which the said Richard incurred, as regards the fishery, damages amounting to £60. Also, when Richard was at Trosselan on the service of the king, the aforesaid fulling mill was broken down and carried away by similar floods, so that the said mill could not be worked for a quarter of a year; by which and such-like common inundations during the said term the said mill often remained wholly without earnings; for which the said Richard has claimed on account of multure and breakages damages to the extent of £54.

III. DEE MILLS, CHESTER.

Richard Ingeniator was mayor of Chester in 1304, but he seems to appear no more in connection with the mills. Ormerod tabulates Robert de Glasham "farmer of the mills and bridge of Chester," 1321; and Bartholomew de Northworkyn "farmer of the mills and fishery of the Dee for three years," 1347; and Pennant states that Edward the Black Prince granted the mills for life to Howell of Favyall, as a reward for bravery at the battle of Poitiers.

3. Claims by Lessee, 1278–1289.
Hist. Cheshire, i. 225.

4. At Chester the year 1356 (29 Ed. III.) was an important one; the king then confirming by royal writ, as the recognised legal customs of Dee Mills, those usages which had already prevailed there time out of mind; usages to which in the numerous law-suits of the lessees of these mills in the seventeenth century their pleas as to custom invariably go back, as to a solid foundation rock of indisputable evidence.

4. Customs and Court, 1356.

There were and are divers ancient customs belonging to the said mills in Scīo Cestriæ temp. Edward III. intituled:—

Harl. MSS., 2081. 168.

Consuetudines Molendinos de Dee vsitat a tempore cuius contrarium memoria non existat: viz.—

Quod omnes manentes in civitate Cestriæ debent molare blada sua in molend. de Dee et debent dno xvj grañ pro tolnet: saving that there are some particular persons excepted who are noīated in the Records.

Custos dño molend sive fermarius sive alius solebat tenere Cur. sua de oibus transgr̃ factis in eisdm̃ molend. sive in piscar adjunc: et nullus attach. in molend pdcis ex quacunq. causa.

Si aliquis cariavit blada sua ad molend aliqua alia molend. prima vice dcā blada erunt forisfacta, et firmarius dicto molend. habit blada pdca: et si quis bis cariavit firmarius habit blada pdca: et dno Comes tertia vice quod sic cariavit cm̃ equis suꝑ quod blada sunt cariat forisfacta.

Nullus dę civitate pdca habit molas manuales in piudice dco molend.

With particular customs for fees and vailes to the milners.

In the Exchequer Court, Chester:—

Customs of the mills of Dee from the time whereof memory to the contrary exists not, viz.:—

Every one dwelling in the city of Chester should grind their corn at the mills of Dee, and pay the lord the sixteenth vessel for toll; except certain persons named in the records [namely, the tenants of the abbey of St. Wybergh].

III. DEE MILLS, CHESTER.

4. Customs and Court, 1356.

The custodian of the mills of the lord, whether a farmer or other, used to hold his Court respecting all transgressions done against the said mills or the adjacent fishery; and no attachment might be served for any cause in the said mills.

If any one carry his corn to grind at any other mill, the first time the said corn shall be forfeited, and the farmer of the mills shall have it; if he carry it away a second time, the farmer shall have it; and the third time the corn with the horse upon which it is carried shall be forfeited to the lord the earl.

No one in the city shall have hand-mills in prejudice of the lord's mills.

With some other particular customs for fees and vailes to the milners.*

To the foregoing citation may be added another giving further details :—

Harl. MSS., 2084. 4.

In order that the city and county might be well served the kings maintained a clerk of the said mills, who was always present and had allowance of vj[d] per diem: and also appointed the Chamberlain of Chester to take the accounts of the said mills and fishings; and the Justice of Chester to keep courts yearly in the said mills for the better ordering and government thereof.

Ibid., 2081. 98.

Under these customs the mills were long conducted. A writ for the holding of a court at Dee Mills, issued 3 Henry IV. (1402), requisitions the assembling of twenty-four jurors, who were to attend under pain of a fine of 6s. 8d. Another writ of about the same period, under the seal of the County Palatine of Chester, directs the mayor and sheriffs of the city to warn a court to be holden at the mills of Dee before the king's justiciary and chamberlain; and to return a jury to enquire as to an alleged offence of a tenant in withdrawing custom from the said mills. The procedure, therefore, was for the city officers, who themselves had no authority within the mills, to issue a notice through the city that a court was to be holden at the mills; and to summon a sufficient number of reputable and worthy citizens to constitute a jury. In this particular case, one, William Dowell, of Chester, was found guilty of carrying forty-one bushels

* These extras, which it would have been interesting to know, probably gave rise to some of the allegations of extortion leviedga ainst the Chester millers at the beginning of the next century: *q.v.*

4. Customs and Court, 1356.

Text, post, p. 115.

of wheat to be ground at Haffod Mill. In 1635 the holding of the court seems to have been allowed to lapse; and the millers of the period found it desirable to take counsel's opinion on the question, "Whether the fermors, having not of late kept courts nor seised corne according to the ancient custom, have lost the same by non-user?" being consoled by the obliging opinion that "Ye non-user doth not destroy ye custome." The salary of sixpence a day paid to the keeper, a fairly good one, compared favourably with three-halfpence a day paid in 1364 at Perth Mill, purchased by Edward III. in 1329 and duly placed in charge of a keeper.* The nature of the duties of a keeper are suggested by the various designations under which he appears—clerk and keeper, clericus apprimator, clerk and approver, &c.; and the list at Chester at times includes the appointment of a royal favourite or body-servant of the sovereign—valettus honorabilissimus, valettus camere; these occasionally being installed for life, and having deputy keepers and other servants appointed under them. An instance of the formal binding of one of these deputies occurs in the year 1409.

5. Clerical Lessees, 1356.

5. The foregoing confirmation of the customs in 1356 took place in connection with a lease of the mills in that year to four partners not hitherto included in the recorded lists of officials of the mills:—

Harl. MSS., 2081. 181.

Indenture of the lease of the Dee Mills, 29 Edward III., by the king to Robert de Bredon, parson of the church of St. Peter in Chester; Simon de Asewell, clerk; John de Newark, clerk; and Nicholas de Eccleston, carpenter: as well fulling mills as corn mills, with fishings on both sides of the bridge, but not the fishing under the bridge when the same shall be finished [the bridge, destroyed in 1227 and 1280, again appearing to have been wrecked], which is reserved to the king; with all customs and multure for three years;

* The accounts of the provost of Perth in the Exchequer Rolls of Scotland for 1364 include "payment made to a certain keeper of the mills during the time they were in the hands of the king, for the custody of the mills and collection of the receipts for the year, 44s. 4d."

III. DEE MILLS, CHESTER.

5. Clerical Lessees, 1356.

paying therefor to the king at his Exchequer in Chester one hundred fourscore and ten pounds yearly. The king granting timber from his woods and forests sufficient, by the view of his foresters or carpenters, for wheels, cogges, rungs, sheldes, ladles, and other necessaries of timber whatsoever for the same mills during the term. The aforesaid Lord Edward at his own proper cost to make anew all the wheels of the aforesaid mills and other necessaries that shall be required; and to find sufficient stones for the mills; and to sustain the mills in great timber, kilns, &c. The lessees to provide all else required in the working at their proper cost; and to deliver up the mills at the end of the term in as good a condition, &c. The king to make [rebuild] and sustain the causeway in the river; and if the foresaid causeway so by him built be damaged by misfortune or rage of water, the lessees not to be chargeable with the damage.

Finally, the king wills that his Justice of Chester or his deputy shall yearly hold the foresaid court in the foresaid mills during the whole time aforesaid and whensoever by the said lessees reasonably required so to do.

In the hands of the three clerics and the carpenter the undertaking seems to have prospered. The mills were kept in constant repair, as is evidenced by various of the king's precepts for timber from Delaware Forest for the purpose. The rent of mills and fishery, which in 1289, as paid by Richard the Engineer, had been £200 per annum, was now, in 1356, reduced to £190. In 1377, however, it was again increased, as appears in the returns of revenue from the entire royal earldom of Chester in that year; the total being £1,001 3s. 7d., towards which the fee-farme of the city contributed £100, and that of "the mills upon the river Dee" £240.

Poole's Cheshire, 1778, ii. 658.

Matthew Haydock was "clerk of the mills, Chester, during the king's pleasure" (1384), and William Mareschal "clerk of the mills of Dee for life" (1392). It appears to have been during his tenure of office that the Black Friars of Chester in 1396 were made "tolfre" and "hoper free" at these mills; a privilege which, it may be added, was extended to the Carmelite Friars of Chester, February 13, 1400, in response to their petition to Henry, Prince of Wales, that they

Hist. Cheshire, i. 225.

Text, III. 47, 63, 66.

were so impoverished by a recent great murrain and a raid committed in the parts round about them that they "could not serve God or honestly live without aid."

6. Alleged Malpractices, 1397.

6. Another new lessee is here to be added to the lists; one, John Walsh, either a most iniquitous or a much-maligned specimen of the millers of the Dee. In 1397 the corporation petitioned the king for a lease of the mills, declaring that a certain "John Walssh, also seeking the lease, purposes to use it in order that he might express his ill-will to this corporation by levying excessive tolls and extortion." Their petition was refused; and Walsh, who may have made a higher bid (as, in view of his alleged contemplated extortion, he might be able to do), duly secured the mills. The allegation of the corporation appears to have proved correct; and in the same year (1397) Richard II. was in receipt of the following humble petition from the city for redress:—

> To th exellent and or most redowbted sovereyne lord the Kyng. Harl. MSS., 2003. 857b.
>
> Youre pore leges and supplycants the meyre and comltie of yor citie of Chester showe that the mylners of yor milnes of dee do take from day to day divers owtragiouse partes over the right toll used throghe all yor realme of ther corne and malt over the xvj greyne after that it is grounde to the great ympoverishment and damage of yor said pore supplicants.
>
> That it mey please yor most highest maiestie and of yor especiall grace to graunt unto theym to have ther corne and malt grounde at yr milnes for the xvj greyne without any other tole parte or any thyng over. For God sake and in love of the citie.

King Richard, Prince of Chester, lost no time in coming to the relief of the "pore leges"; ordering on February 23, 1398, that no more than the accustomed sixteenth rate of toll be taken for a period of four years; this term probably completing the duration of Walsh's lease. But before the termination of the period Walsh had vacated the place, for in 1400 other keepers are found in possession. Further, he did not settle his accounts when leaving, possibly owing to his having Ibid., 2003. 859.

III. DEE MILLS, CHESTER.

6. Alleged Malpractices, 1397. Harl. MSS., 2057.

pending one of the usual claims for repairs, &c., against the king; and in 1405 a writ issued by Henry IV. ordered "John Wales and Mary his wife, late fermors and receivers of the profits of mills of Dee," to appear before the sheriffs and answer for their arrears in respect of the same. Despite the foregoing invidious record, John Walsh seems to have been a man of repute in Chester. He was mayor in 1407, and again in 1411; while in 1409 he appears as one of the dozen sureties for R. del Hope, a deputy keeper of the mills; and it seems, in fact, probable that the extortions of which he was held to be guilty were based on customs he had found in operation; and which, indeed, were enforced after he left, with important results.

7. Inquisition, 1400.

7. Thomas de Mostyn was the next appointed lessee, and during his administration a series of alleged extortions and abuses, said to have been rampant "for forty years and more," reached an acute stage. The citizens seem to have rebelled and demanded an enquiry; in any case an inquisition was held, at which amazing evidences of malpractices by the millers were produced. According to the finding of the jury, Mostyn and his subordinates, like their predecessors, were guilty of extortions alike against bakers and burgesses, brewers and "paupores"; and so important was the verdict esteemed with regard to the true customs, usages, and tolls of the mills, that as late as the reign of Queen Elizabeth it was rendered into English and entered at length in the Pentice Chartulary of Chester.* The following transcript is taken from this record, still existing among the muniments of the city, and collated with a copy of the same by Randle Holmes in the Harleian MSS.

Harl. MSS., 2083. 517.

* The Pentice, or Penthouse, was the court-house where the mayor anciently dispensed justice.

7. Inquisition, 1400.

The new names of officials added to the lists by the document are those of Thomas of Mostyn, official "clerk and keeper"; William Brooke, "striker," who measured the grain and flour, sweeping the top of the filled bushel level with his "strike," so that it should be "razed" and not "upheaped"; John of Burton and David of Mylward, the actual operative millers. One of the events alluded to, an alleged extortion practised upon some men who brought grain in a boat from Wirral—which, as will be seen, was not an extortion at all—is stated to have taken place in the reign of Richard II., and apparently was thus really referable to John Walsh; but the others are definitely stated to refer to Thomas Mostyn and his subordinates. The customs, it will be noted, are several times dated back to the time of "Rannulphe late Earl of Chester," apparently Randolph Blundeville, sixth Norman earl, who died in 1232; the period of the inquisition appearing to be the early part of the year 1400.

Pentice Chartulary, Chester.

Ordinance at the Dee Mills for the citizens:—

Ancient toll.

The Jury say by their oath that where the citizens of the city of Chester in the time of Rannulphe late Earl of Chester were houlde to grind their corn at the milnes of the said Earl, of the Dee, by sixteen graynes of every . . . of corn:

[One-sixteenth of the total quantity, the average toll throughout the kingdom.] Text, III. 154.

Extortionate toll.

Now Thomas of Moston, clerke and keeper of the said mills, William Brooke, striker, and John of Burton and David of Mylward, milners there, doe take and of long tyme have taken by extortion, to the great damage of all people, at the feaste of the birthe of our lord god, of everye man in the said cittie dwellinge

unam tribulatam d tribulat et quartam ptem unum tribulat farine que dicit schole, de quibusdam tribulat tres facient unum hepe non emulat, et tantum farine ultra hoc quantum supra manum suam extensam capere poterint; et insuper capint de quolibet paupore quod unum bs vel d bs bladi ad dicta molendina molare tantum quantum de quolibet . . . gerent ibm quatuor bs vel sex bladi:—

one measure, half a measure, and a fourth part of a measure, which [entire quantity] is called a schole; three of which measures make one hoop not heaped up; also as much flour beyond that

III.
DEE MILLS,
CHESTER.

7. Inquisition,
1400.

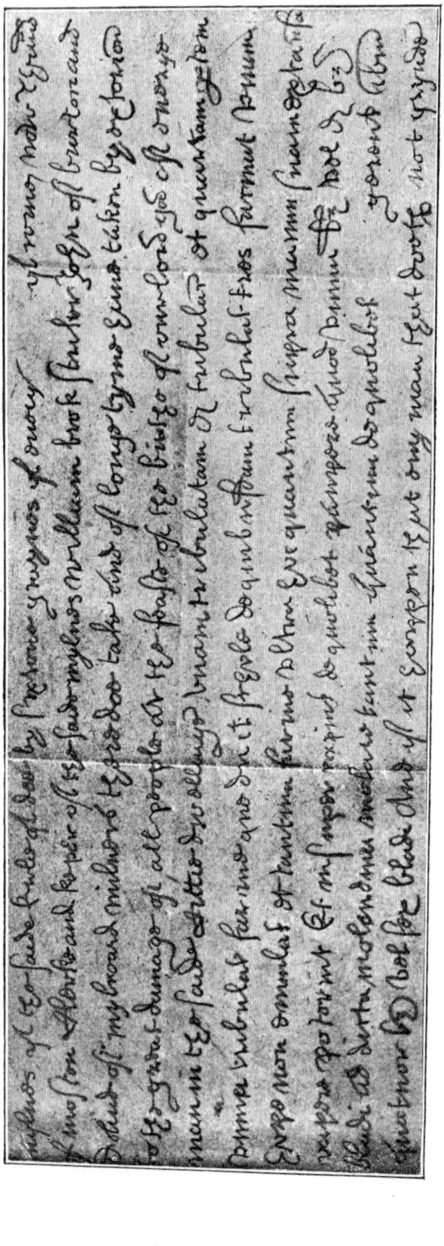

INQUISITION, 1400: EXTRACT FROM THE CHESTER PENTICE TRANSCRIPT: *temp. Elizabeth.*

quantity as might be held on the extended hand; and, besides, they take from each poor person who brings one bushel or half a bushel of corn to the said mill to be ground as much toll as should serve for four or six bushels carried there;

7. Inquisition, 1400.

And if it happen that any man as doth not grind his corne afore the feaste aforesaid at the said milnes, and, after the said feast doth go to the said milnes with his corn there to grind, the aforesaid keeper and milners do take of every such man in the same manner and form as they do take afore the said feaste: that is to say, the said measure called a schole: In great extortion of the common people, which is worth by year by estimation x mks [£6 13s. 4d.] and more: And the said extortion they and their predecessors have continued for xl years and more.

[The ambiguity or illegibility of the first part of this clause, which has constrained the Elizabethan transcriber not to attempt its translation, still renders its meaning slightly obscure.

The schole comprising $1\frac{3}{4}$ measures is again referred to in the clause with regard to the bakers; where its true size is stated to be one span in length and one in breadth. This illegal schole was $1\frac{3}{4}$ of $\frac{1}{3}$ of a hoop = $\frac{7}{12}$ hoop, true measure and razed. By the clause "Too Large a Toll Hoop" the true hoop seems to have been half a peck; on which basis the illegal schole was equal to $\frac{7}{24}$ peck. Text, III. 162.

The tribulus or toll-dish and the hoop-measure are mentioned in the inventory of Dublin Mills in 1319, and appear in various medieval records. Ibid., ante, p. 22.

The taking of flour by the handful, illegally practised here, was more anciently a legal method of rough measurement.] Ibid., III. 148.

For the bruers.

They say also that where bruers in the said city were wont to grind their malt in the milnes aforesaid at sextene graines as is above said, that is to say, of sexe sacks or eight sacks, one tolle hope to pay; now, the present milners of every sack containing twoe cranocke. . . . Which tolle hope they were used to have of the aforesaid bruers to their own use for their sustenance, so that they should help the said bruers in the said milnes to bear their corne to the hopper to be ground, and to lift their sacks upon their horses: Now they will not help as they be bounden: and the aforesaid keeper and milners do take and of long time have taken of every six graynes one tollhope in extortion: And above that tollhope do take half a tollhope and the fourth part of a tollhope where they ought to take nothing: To the great damage and extortion of the common people to the value by year of twenty pounds and more, and the same extortion they and their predecessors have used of forty years as is aforesaid.

[The matter of assisting to load and unload the brewers' horses, either for a fee or without, was precisely one of those customs established by usage, if not set out in original customs rolls, liable to abuse. It is to such a detail as this that the extra "certain fees and vailes to the milners," set out in the Ancient Customs, would apply. If such fees were once allowed to lapse, any resumption of them would inevitably be declared an extortion.] Ibid, III. 133

For money paid at the milnes.

They also say that in the time of the said Earl Rannulph it was made an order that every man going to the said milnes with his corn there to grind that only he should pay but his toll, without more: Now the aforesaid keeper and milners do take above the said toll of divers men,

III. DEE MILLS, CHESTER.

7. Inquisition, 1400.

without number, a penny, two pence, sometimes four pence, sometimes five pence or six pence to their own profit; which money they divide amongst them; as well of men in the city dwelling as in the country abiding: Or else their corn in the said milnes they cannot grind by the said keeper and milners: Or, other, the said corn within the said milnes shall tarry unto they have satisfied the said keeper and milners as is aforesaid: To the grievous damages and extortion of the common people of the said Earl; Which is worth by year a hundred shillings and more: And so that extortion hath continued of tenne years and more.

[This extra charge appears to be the same stated in a later clause to have been levied on the men bringing corn from Wirral by boat, viz. a rate above the one-sixteenth grist the millers were entitled to charge citizens. The complaint was that the millers not only took money (grist toll only being legal), but, further, that they surcharged citizens, as though "foreigners" from outside the city.]

What men should and ought to be in the milnes.

They say also that in the said milnes ought to be five maisters and one maister above them with his knave: And now they be but three with their knaves, which also take their part after the rate of his porcion: And the said keeper has taken the same parts [for the men not engaged] to his own use and profit, for and in the name of the said four maisters [and their knaves] for the rate of their portions: In prejudice and damage of our lord the King and Earl, and in extortion of the common people: Which is worth by year one hundred shillings and more: And the same hath continued by eight years and more.

Text, III. 130.

[This cause of complaint has already incidentally been explained.]

How the bakers ought to grind.

They say also that the officers in the said milnes, which now be, and all their predecessors which afore this time there have been, in the time of the said Earl of Chester, were wont to take of every baker of the City of Chester only (and of none other citizen of the said city:) That is to say of every xiij bushels of corn one schole full of meal; Which ought to contain in length one spanne, and in breadth one other spanne, without more: And now they take, and of long time have taken, of every six bushels of corn, in extortion and above sixteen grains, one measure and a half and the fourth part of a measure: Which said measure is larger and more than the measure which was ordained in the time of the said Rannulphe Earl of Chester: Which is called a schole by a great quantity [of people], that is to say one schole and a half and three parts and more: To the great damage and extortion of the people of our sovereign lord the King and Earl of Chester, to the yearly value of foure pounds and more: And so they and their predecessors have continued for forty years and more.

Ibid., III. 112.

[Bakers, like other citizens, were bound to grind at Dee Mills. The Charter of Incorporation granted to the company in 1463 also astricted the members to the mills. The toll payable by bakers "and by none other citizens of the same city" was a special rate for the trade. Such a special rate was payable by the bakers of Paris in 1258.]

7. Inquisition, 1400.

How bruers ought to grind.

They say also that if any brewer, or his servant, to the aforesaid milnes going with their malt there to be ground, whether it be much or little, if by chance any of the said malt do fall upon the ground beside the milne where it is ground, although it were within the Rynde or without, the aforesaid breuer nor his servant shall not be bound to take the said malt to his house, although it were ground or not, for fear of the said milners: In great extortion of the common people: Which the said keeper and milners have taken to their proper uses and not to the profit of the lord: by year by estimation ten pounds: And so they and their predecessors have continued for forty years and more.

Text, III. 164.

[Grain within the rynd was actually within the millstone, whence it was almost impossible for it to fall, unless improperly fed from the hopper; the special accusation being, however, generally that grain falling on the floor of the mill became the perquisite of the miller.]

For the schole of the milnes.

They say that by common assent of the said milners and keepers they have a certain scole that they take in the name of the said Earl with that [besides that] they take to his commoditie and profit: That is, against the right of the said lord, sometimes for five pence, sometimes for four pence, and sometimes more and sometimes less: Which said money they divide among them: In prejudice and deceit of our sovereign lord the King and Earl of Chester, and to the great damage and extortion of the common people, of the yearly value by estimation of forty shillings: And so they and their predecessors have continued of forty years as is aforesaid.

Grinding toll corn.

They say also that where people of our sovereign lord the King and the Earl of Chester do buy corn at the said milnes called Tolle Corne, by which all men that do buy the same shall be free without any toll [for grinding] therefor paying: Now the aforesaid milners do constrain and do take toll of the said corn so bought: In extortion of the common people by the yearly value of ten shillings.

Hollerin.

They say also that there is a certain payment in the said mylnes called Hollerin which he doth take of every sack a farthing where they ought to take no farthing: Which said farthing the said keeper and milners do take to their own proper uses: In extortion and great damage of the common people, to the yearly value of forty shillings: And so they and their predecessors have continued by forty years and more.

For malt grinding.

They say also that if any man or woman or their servant, to the said milnes coming with their malt there to grind, the foresaid milners the same malt will not grind [properly]: but a great part thereof shall be whole, and not well ground nor truly, unto he pay the said milners a penny or twopence for grinding of the same malt: otherwise the said malt shall not be ground in default of the said milners: To the grievous damage of the common people, to the yearly value of ten shillings: And so they have continued by ten years.

Too large a toll-hoop.

They say also that where the said keeper and milners have used to have a measure called a toll-hope in

III. DEE MILLS, CHESTER.

7. Inquisition, 1400.

the time of Ranulph Earl of Chester, of which toll-hope sixteen make a crannock: Now the aforesaid keeper and milners have, and of long time have had, another measure called a Tolle-Hoope of the which Tolle-Hoope thirteen make a crannock: To the great damage and extortion of the common people, to the yearly value of forty shillings: And so continued by ten years.

Text, ante, p. 18.

[The true toll-hoop is said to be the one-sixteenth of a crannoc. According to the calculation at Dublin in 1320, the crannoc was equal to eight pecks; the capacity of the Chester toll-hoop being therefore half a peck. But little reliance is to be placed on these varying local measures of capacity.]

Grinding for the poor.

They say also that where any man or woman doth bring three or four bushels to the said mylnes to grind, the milners aforesaid do take of every sack two handfuls, or three, at the least, to the use of the said keepers and milners whereof which they ought to take nothing: In extortion of the common people, and worth by the year twenty shillings and more: And so they have continued by ten years and more.

Ibid., post, p. 100.

[This complaint claims a custom of grinding small quantities, for the poor, toll free; and two centuries later we really find such a custom definitely stated at an inquisition regarding Dee Mills: "The inhabitants of the said city do for the most part grind their corn and malt at the said mills with great ease to themselves, especially the poorer sort, who usually grind their pecks and small portions of corn without paying any toll for the same."]

How men of Wirrall came to grind at the mylnes.

Thy say also that where men did come of the parts of Wirhall by water in a little boat with a certain quantity of grain, to the said mylnes to grind, against the purification of our blessed ladie [February 2] in the . . . year of the reign of King Richard the Second, then came the said milners, by the assent of the said keeper, and made the said five men to swear upon the holy evangels within their hands, how many men were partners in the said corn: It was admitted by them that fourteen were partners; and upon that they took of every man a penny, afore they would grind, besides the toll; In extortion and grievous damage of the common people, to the value of fourteen pence.

[Wirral is the peninsular portion of Cheshire lying between the Dee and the Mersey. The charge of one penny for each man owning a share in the little boat's cargo, in addition to the usual toll, may be evidence of the sharp business qualities of the millers, but cannot be alleged as an abuse of the legal customs of the mill, for the Wirral men were not citizens of Chester, to whom alone the customs applied; and, in fact, these adventurers were doubtless absenting themselves from their own local mill, as the Dee millers very readily perceived, and were surcharged as all other than citizens might be, at the discretion of the millers.]

8. Installation of Keeper, 1409.

8. The foregoing tremendous verdict wrought the desired effect. Thomas de Mostyn and his satellites promptly disappeared before the close of the year 1400, Henry de Strangways in that year being "clerk and keeper of the mills of Dee for life"; being succeeded, June 9, 1401, by Robert Castell, gentleman, as "clerk and approver." Under him, as deputy, served John,

son of William del Lowe (1407), also John de Huxley and Robert del Hope (1409). In this latter year Hope, under the guarantee of twelve sureties, was duly installed as the actual practical manager of the mills :—

III. DEE MILLS, CHESTER.

8. Installation of Keeper, 1409.

Coppie of a Recognizaunc wherbye sertaine are bound for the occupier of Dee mylls 23. Oct. 11 H. 4.

Harl. MSS. 2081. 179.

Re Robti del hope deputati Rbt Castell clici & Apprimatoris Molendinor de Dee.

Johēs Wallssh de Cestr', Ranulphus le clerke de Cestr', Johēs del hope de Cestr', Rogerus le potter, Ith Trevor, Johēs de hale de Cestr' Vyntnyer, Innocencius de Chesterfeld, Thomas Aleyn, Ricūs Strangways, Johēs Launcelyn, Thomas de Sherdeley & Thomas de Gorstylowe venerunt in Sc̄cio hic xxiij° die Octobris Anno reg. henr' quarti post conquestum vndecimo & manuceper' et q'lbt eor. ꝑ se manucepit ꝑ Robto del hope deputato Robti castell Clici et Apprimatoris Molendinor' dni' henr' principis Wall' & Comitis Cestr' de Dee & piscarie eiusdem principis & Comitis ibm̄ de bene & fidelit' s'uiendo erga dco. dno. Comiti sub Cam'ar' Cestr' ꝑ tempore existente de p'fato Robto Castell in officio ꝑdco qm̄diu in eodem officio sub-eodem Robto Castell stetit & se esse contigit: Et de se bene gerendo erga dco. dno. Principem & Comitem & omēs sectatores Molendinis ꝑ'dcis: Et ad bene et fidelit' soluend dco. dno. principi et ꝑfato Cam'ar' vel eius locum tenen' ꝑ tempore existen' noiē et ad opus ipsius dni' principis quodcumq' denarios de exitib' Molendinor et piscarie ꝑdcōr ꝑuenientes de tempore in temꝑris ad mandatu' ipsius Cam'arij vel eius locum tenen': Et eidem Cam'ario vel eius locum tenenti intendens et obediens erit tam in solucoē denarios ipsius dni' principis ac vendicoē granor brasr et pisc'm de molendin' & piscar' ꝑdcis ꝑueniencm̄ et illa sub consilio eor̄dem vendet, qm̄ in om'ibꝫ alijs officio ꝑdco qualit'cumq' spectan': Et ad om̄es et sing'las leges & consuetudines in eisdem Molendinis hactenus vsitatas ꝑ posse suo observabit ad opus et ꝑficium ꝑdci' dni' principis & Comitis subpena que incumbit, qm̄diu idem Robtus del hope in officio ꝑdco esse contigit.

Recognisances for the occupier of Dee Mills, October 23, 1409.

Re Robert del Hope, deputy of Robert Castell, clerk and conductor of Dee Mills.

John Walsh of Chester [with eleven others named] have come to this Court of Exchequer, October 23, the eleventh year of the reign of Henry, the Fourth after the Conquest, and bound themselves conjointly and severally on behalf of Robert del Hope, the deputy of Robert Castell, clerk and manager of the mills of the Lord Henry Prince of Wales and Earl of Chester on the Dee, and the fishery of the same prince and earl there, for his good and faithful service towards the said lord the earl, under the Chamberlain of Chester, for the time the said Robert Castell continues in the said office, so

III. DEE MILLS, CHESTER.

8. Installation of Keeper, 1409.

long as the said R. del Hope may remain in the aforesaid office under him: also for his good conduct towards the lord the prince and earl and all the suitors of the said mills: Also for his true and faithful payment to the lord the prince and the aforesaid chamberlain, or whoso may be named as his deputy for the time being, to the use of the said lord the prince, of all moneys arising from the said mills and fishery from time to time, as directed by the said chamberlain or his deputy: Also that he will be attentive and obedient to the said chamberlain or his deputy for the time being, as well in the payment of the moneys of the lord the prince, as the sale of grain, malt, and fish derived from the said mills and fishery, selling the same subject to their counsel, as in all other matters whatsoever concerning the said office: Also that all and singular laws and customs of the said mills hitherto used he will observe to the best of his power to the use and behoof of the said lord the prince and earl, under the penalty provided, so long as the said Robert del Hope may hold the said office.

9. Bakers' Company bound, 1464.

9. The bakers of Chester, being citizens, were of course personally bound by the customs of the Dee Mills to grind there; still, on their incorporation the astriction of the company was formally made a condition of their constitution. In 1463 the mayor had issued a charter which, duly recognising the bakers' rights and privileges, had ignored their responsibilities to Dee Mills; but in the year following, when this was presented to Edward IV. for confirmation, the king "let them to wit" that astriction to Dee Mills was not to be overlooked, and inserted a clause to that effect in his confirmation.

Harl. MSS., 2054. 41[b].

Be it knowne to all men that in the citty of Chester ther is and hath byn out of tyme of mynd a craft called the craft of comon bakers of the same citty; having belonginge thereto an alderman and two stuards to be chosen by the brethern of the same craft: Which byn and have byn tyme out of mynd bounden to make such p'vision that there be contynuall brede sufficient for alle the kings people beinge there at alle tymes:

And in the same citty of Chester it hath ben uset, out of tyme that noe mynde ronnes, that noe mannr of man shuld use the craft of a baker in the sayd citie, nor noe brede bake therein to sell but he were addmited by the alderman and the stuards of the same craft for the tyme being to be one of there brethren in that occupacion; and have and deliver up his marke [stamp for bread] in wax before the maior of the said citty for the tyme beinge;

9. Bakers' Company bound, 1464.

and there abyde;
and be sworne to be ruled in alle thynges concerninge the same craft by the alderman and the stuards thereof for the tyme beinge;
and be redy to pay for the costes and expenses of the Play and Light of Corpus Christi as oft times as it shall be asseset by the same stuards for the tyme beinge;
and pay unto the same alderman and stuards for the tyme beinge, if he be a forener that shall be so admitted xl^{s}, and if he have byn aprentice to the same craft within the same citie xxvj^{s} viij^{d}, and a dinner to the brethren of the same craft [on his] firstcomminge thereto, with all other costes and charges belonginge to the same craft;
and keep such assiz [of bread] as shall by the maior of the same citie wekely be assigned unto the same stuards for the tyme beinge.

And it is so now, that dyvers ill disposed psons of the same citie, not admited unto the sayd craft nor sworne in forme aforesayd, in late tyme have taken and presumed upon theym, and daly yet usurpen and presumen, to bake brede and it to sell within the same citty; not keepinge any assize nor doyinge nor payinge any other duty as is aforesayd; Which brede so baken for the most parte is not sufficiente nor able; to the grete and generall hurte and desceyt of all the kings people, and greate disclaunder of the maior of the sayd citty for the tyme beinge, and of all the craft of bakers in the same citte, and contrary to the custome aforesayd; and like to groe to a fynall destruction of the same craft;

Wherefore, we John Davidson, alderman; Robt. Ranson and Thos. Hobbs, stuards of the same craft by the assent of alle our brethern of the sayd craft, and by the advice, goode wil and assent of Robt. brunne, maier of the cittie of Chester, and his counsell;

Consideringe the sayd custome and the mischeyf and inconvenyence that have faln and likely so to fall by cause of the usurpation and brech of the sayd custome;

ffor the Tranquility and comyn weale of alle the kings people, and [for ceasing] the sclaunder aforsayde, and affermynge the sayd custome to be goode;

Wil and ordayne that noe manr of man, of what condicion soo ever he be, occupie the craft of baker in the sayd citie, but if he be admited and sworne in forme aforesayde, upon payne of liij^{s} iiij^{d} to be forfet by them that do the contrary to the custome aforesayde as oft tymes as any pson psumes soe to doe: The wan halph of the same payne to be forfet and payde to the kinge and his heyers, Erles of Chester; and the tother halph to the alderman, stewards and brether of the sayd craft for the tyme beinge: The same payne to be levied by the sherifs of the sayd citie and the stuards of the same craft for the tyme beinge, by such meanes as shall be thought most expedient by the same sherifs and stewards for the most hasty levy therof:

Also whereas also it hath byn uset out of tyme of mynde that if any brede baked out of the sayd citty be brought to sell within

III. DEE MILLS, CHESTER.

the same cittie upon any market day in the same citie, that such brede as remayneth thereof unsould the same market day noe pson of the sayd citty shuld bye any such brede to sell agayne but on a market day:

9. Bakers' Company bound, 1464.

We the sayd maior and his counsell ordayne that noo mannr of psons of the sayd citty put to salc nor bye within the same citty any such brede baken out of the sayd citty only on the Wennysday or Seturday, which be market days uset in the same cittee, agaynst the custome aforsayd upon payne of xx^{s} [to be levied as before stated].

In Wittnesse wherof the sayd maior to this psent writinge hath set the seall of his office; and the sayd alderman and stuards of the same craft with alle there brethrn of the same craft in the same citie—that is to wyt, Jo. dyson, Jo. lancastr, Robt. botyll, Danall baxter, Jo. Corroke, William Masey, Thos. Mason, Rich. Anyon, Tho. Coke, Willm. Jellemyn, Randolf baxter and Edwd. dyson—to this writinge have set their seales.

Geven at Chester the xii day of January, in the yeare of the reigne of Kinge Edward, the fourth after the Conqueste, the seconde [1463].

Harl. MSS., 2054. 45.

Edward, &c. To our trustie and welbeloved the maier and sherriffes of our cittie of Chester.

We lete you to wite that we have understood that by a supplicacon presented to us on behalfe of our welbeloved the alderman, stewards and brethren of the craft of bakers within our cittie of Chester, that Wheras it hath been ordeined and used tyme out of mynd [&c., as in the above civic charter of 1463] . . .

And that the bakers soe admitted owe their sute unto our milnes there called dee milnes and be bounden to grind their cornes there and in noe other place:

And whereas certain people are using the craft without being admitted as bakers [&c., as in the charter].

Geven at Westminster under our Privy Seal 2 June, the third y^{r} of our reign [1464].

Ibid., 2054. 42.

The above was confirmed by an inspeximus charter of Edward, Prince of Wales and Earl of Chester, 29 November, 18 Edward IV. (1479), four years before his murder, as Edward V., in the Tower of London. Seventeen years later a confirmation was granted by Prince Arthur:—

Ibid., 2054. 42.

The most illustrious Arthur, eldest son of Henry the Seventh, King of England and France and Lord of Ireland; Prince of Wales, Duke of Cornwall and Earl of Chester, to all to whom these letters shall come, greeting:—

We have inspected a certain ordinance of the alderman, stewards and brethren of the craft of bakers in our city of Chester framed for the convenience and benefit of the said craft and of our mills there in these words:—

"Be it knowne to all men that in the city of Chester ther is and hath byn out of tyme of mynd a craft called the craft of comon bakers" [&c., reciting the original of January 12, 1463].

Which ordinance by the deliberation and advice of our councillors we approve and confirm.

9. Bakers' Company bound, 1464.

Witnessed by myself at Chester 18 Decr. the 11th yr of the reign of our said father [1496].

A second civic charter, dated April 26, 1553, was confirmed by Edward VI. July 6, 1553*:—

Harl. MSS., 2054. 44.

To all people unto whom this presente writtinge shall come to be seene or herd, Mr. William Glassier, gent., mayre of the cittie of Chester; Rich. Sneyd, gent., recorder of the same citie; and the aldermen and justices of the peace within the county of the citty of Chester, whos names be underwritten, greetinge in the lord god everlastinge:—

Know ye that whereas we the said mayre, recorder and aldermen; together with Willm. fflecher, baker [alderman], and John ffayreclyf and John Smith, stewards of the mystery or craft of bakers within the sayd citty of Chester, and the brethren of the sayd crafte; and at the present contemplation of the sayd alderman and stewards of the sayd mystery, and the brethren of the sayd crafte:—

Have perfect knowledge and clere understandinge that

"Wheras within the sayd citty of Chester there was and hath byn tyme out of mynd a crafte called the crafte of comon bakers" [&c., reciting the charter of January 12, 1463; but amending it in one or two passages to suit the modern times. For example, in the first clause the words "good and holesome" are inserted before "brede"; and the compulsion to pay towards the expenses of Corpus Christi play is omitted.]

As by divers ordinances of ancient time, agreements of the craft with the mayor, divers charters and letters patent doth more plainly appear: all which aunciente customes and laudable ordinances have been duley perused and examined and likewise discretely pondered, consydered and wayed by the sayd mayor, recorder and aldermen. . . .

Who doe confirme, ratyfy and by these presents approbate, establish and allow all and singular the ordinances aforesayd to be good and profitable; and to stand and take their effects in every parte thereof; and shall be fyrmly continually confirmed, obayed, fullfilled and kept in every point, clause and article within the said citty. In witness whereof, &c.

Given at the Pentice of the said citty the 26th day of April, in the 6th yr of the reign of Edwd. VI. [1553].

per me, WILLIAM GLASSOR,
Maiorem cvtis Cestriæ.

There hath byn 22 seales of which no wax bits remayneth.

* Further confirmations were granted 6 December, 7 Elizabeth (1565), and 30 April, 2 James I. (1605). A Harleian note states, "the bakers' charter is inrolled in the pentice court booke for ye yeare 1569"; and there it remained a standing enactment, binding the trade for a further couple of centuries.

III. DEE MILLS, CHESTER.

10. An Uneventful Period, 1409–1532.

Hist. Cheshire, i. 225.

10. During the fifteenth century the records seem to be very meagre, so much so that the entire period 1409–1532 can but at present be bridged over from the list of keepers tabulated by Ormerod, with what trifling record may be attached to each. In 1420–21 William del Moeles, former clerk and manager of Dee Mills, is declared to owe of his old arrears £4 16s. 9½d.; the farmers of the fulling mills on the opposite side of the river, in the same year, Richard de Hale and his associates, owing also £10 13s. 4d. In the year 1429 Thomas Butler, yeoman, who had had charge of the establishment, was accused of "divers transgressions," and, with all his servants, was ordered to appear before the Court to answer for the same. The staff, as named on September 1, 1429, comprised Stephen Mulner and John Hicson, millers; Agnes, servant; John Bradley, labourer; Thomas Manchester, marshal; William Freget, sporior. But though summoned no fewer than four times in 1430, the parties did not appear; and it seems probable that the matter in dispute was soon after settled, the only indication of the fact then remaining being the negative evidence of the disappearance of their names from the rolls.* On May 2, 1436, Thomas de Pulford, "valet of our chamber" (afterwards "of our crown"), who had been appointed clerk and keeper by Henry VI. during the royal pleasure, was reappointed for life. Towards the close of the same year William Foster is mentioned in the Ministers' Accounts as clerk, apparently as the deputy of Pulford; and in 1461 Richard Bold, protonotary, occupied the same subsidiary position. Pulford seems to have held the supervision of the

Pl. Ro., 7-8 Hy. VI.

* No opportunity has occurred for a very close investigation of the rolls; those for the two years under notice comprising over fifty skins of parchment each, many of them over a yard in length, closely written. But possibly the search might prove interesting to any local antiquary with sufficient time and zeal to undertake it. The citations above mentioned appear on m. 7, m. 14, m. 19, m. 23.

10. An Uneventful Period, 1409–1532.

mills for twenty-eight years, and to have been succeeded by Randal [Richard?] Bold, whose appointment was dated by Edward IV. from the first day of his reign, March 4, 1461, and made for life. Bold was living on August 1, 1464, but was succeeded on or before November 4, 1465, by David and William Malpas. The former was "one of the valets of our crown," and the latter "one of the valets of our chamber"; and the appointment was made conjointly for life, with reversion to the survivor of either. This arrangement came to an end in 1476, when William Butler was appointed clerk by Edward IV. during pleasure; to be reappointed for life by Richard III. on his accession in 1483, and displaced by Henry VII., who, soon after attaining to the throne, appointed on May 20, 1485, Richard Gough, and on November 30 of the same year Robert Hanbury, gentleman, for life. Hanbury worked the mills by practical men, as had Robert Castell by R. del Hope. In 1504 Hugh Hurleston, as "farmer of the Dee Mills and the king's pools," is entered in the compotus of the king's receiver for Cheshire as paying on a twenty years' lease £50 per annum for the mills and £24 for the pools. Five years later, July 6, 1509, the successors of Hanbury were William Poole, a member of the king's guards (unus valectorum honorabilissimi gardi nostri), appointed, together with Edward Ap John, during pleasure; these being followed by Hugh Goldsmith; on whose death Urian Dymok, one of the sergeants of the city, was installed, March 19, 1524.

Harl. MSS., 2083. 511.

Hist. Cheshire, Ormerod, i. 225.

During this barren series of years the mills, as will be seen, were directed by non-practical men—soldiers, Court favourites, or local notables—invested with the custody of the establishment at the pleasure of the king or of the Chamberlain of Chester on his behalf. So long as this state of affairs lasted matters

III. DEE MILLS, CHESTER.

10. An Uneventful Period, 1409–1532.

seem to have progressed but indifferently. The deputies, practical workers at the mills, enforced their full toll rights, if they did no more; and there is little evidence of their making any attempt to keep the soke together. Absenteeism became common; and the annual value of mills and fishery, which had been £200 in 1289, £190 in 1356, and £240 in 1377, had decreased in 1504 to the sum of £74 paid by Hurleston. The fact was that vicarious management, such as the official clerks and keepers afforded, was insufficient to cope with the increasing difficulty of maintaining the ancient soke rights intact.

11. A New Era, 1532–1553.

11. This seems to have been the conclusion of the advisers of Henry VIII., who, on April 27, 1532, "devised the said mills, with the multure and appurtenances, to Robert Brooke, esq., for [21] years, who afterwards assigned his interest in the same to Raphe and Thomas Goodman." At this point opened a new era in the management of the mills. The Goodmans, like their heirs and successors, were practical business men, who went to work with a will to restore the falling fortunes of the place. The mills were enlarged and their resources greatly developed; and the harassing labour of building up anew the rapidly disintegrating soke rights in Chester was undertaken with vigour, and prosecuted relentlessly and at all costs, in numerous and protracted actions at law. One result of these ceaseless contentions was to produce a quantity of voluminous documentary evidences; many of which, still existing, render the history of Dee Mills in the sixteenth and seventeenth centuries infinitely richer in detail of law and custom than that of any other of the large mills of the kingdom of which any records appear to remain.

Harl. MSS., 2081. 169.

The Goodmans, whose new régime soon made the welkin ring, had first appeared at the mills in

19 Henry VII. (1504), when, in the compotus already mentioned, Hamon Goodman, with William Shawe and Roger Smyth, "farme ye fulling mille at xj[li] per an." In due course Henry VIII., in 1532, as stated above, demising the corn mills to Robert Brooke, esq., for twenty-one years, the latter assigned his interest in them to Raphe and Thomas Goodman. They at

III. DEE MILLS, CHESTER.

11. A New Era, 1532–1553.

Harl. MSS., 2083. 511.

Dee Mills.—*From the Harleian MSS.*

once heralded their management by bringing an action against Thomas Thorneton, alderman of Chester:—"On August 14, 24 Henry VIII. [1533], divers persons were bound for Thomas Thorneton that he should not carry any corn to be ground at any mills out of the franchises of Chester"; while in 1539 Richard Anyon and several others were similarly bound in recognisances that they should not "carrie cornes to any forren mylle." In the year 1553 Brooke's period of

Ibid., 2083. 520, 521.

11. A New Era, 1532–1553.

Harl. MSS., 2081. 190.

twenty-one years expired; and on June 24 of that year Edward VI. disposed of "the said milnes, with the appurtenances and other things thereto belonging in fee, with all profitts as fullie as the said Robert Brooke or anie other owner or farmer thereof had had or held," to Sir Richard Cotton, comptroller of his household, and Dame Jane, his wife, in exchange for certain lands in Leicestershire. The rental had by this time been raised, under the energetic management of the Goodmans, from the £74 of 1504 to £112 15s. 2d.; and Sir Richard Cotton, on obtaining possession, seems to have promptly renewed the Goodman lease.

12. Demise of Tithes, 1553.

12. Before the close of the year 1553, and because of the events comprised within the few preceding months, the dean and chapter of the new cathedral of St. Werburgh acquired the tithes of the mills. These, which had originally been granted to the abbey by Earl Richard in 1119, being then worth £5 per annum, were now leased to Edward Plankney, of Chester, at the valuation of £9 per annum. The lease, dated 2 December, 6 Edward VI.,—

Cheshire Sheaf, May 1883.

Witnesseth that the said Dean and Chapter have demised, granted, and to farme letten to the said Edward Plankney of Chester, gent., as well all and all manner of tythes of all kinds of grane, corn, and malt to the said Dean and Chapter in and out and from the our said sovereign lorde the king's milnes in Chester, commonly called Dee Mills: as also all and all manner their tythes of fish in the king's pool in the water of Dee: to have and to hold the said tythes of all kinds of grane, corn, malt, and fish . . . for three score years at £9 per an.: that is to say, for the said tithes of all kind of granes, cornes, and malte of Dee Mills £7, and for the tithes of fish 40/-.

13. Action against the Abbey Mills, 1567.

13. The mills at Northgate, which till the Dissolution had belonged to the abbey, had since been worked for some time by Thomas Bavand, who had been sheriff of Chester in the same year as Ralph Goodman had been mayor (1547). Bavand had died and left his widow dependent on the old abbey mill for a living, and she accordingly carried on the business,

13. Action against the Abbey Mills, 1567.

keeping the old connection together. Included in the latter were some citizens who did not live in Northgate Street (the special district for that mill), and who should have therefore ground at Dee Mills. As was afterwards pleaded, the Goodmans had made no demur to this while Bavand lived; but in any case after his death they took action against the widow, as well as several other persons who were working mills and attracting custom that should have gone to the still so-called "king's mills." This suit endured four years.

Harl. MSS., 2081. 109.

In 1567 proceedings commenced with a Bill of Indictment entered in the Exchequer Court at Chester by Ralph and Thomas Goodman against Margaret Bavand, widow, occupier of one watermill, called the Bache Mill, without the Northgate Street [also of the windmill there]; Robert Dandrey, esq., occupier of a windmill near Spittle, Boughton; Thomas Ball, occupier of a windmill at Christleton; John Moreton, occupying a watermill at Great Borrow; Philip Prince, occupying a watermill at Marford:—each of whom, it was alleged, ground corn taken out of the city, to the prejudice of Dee Mills. The plaintiffs pleaded that the late King Edward and his progenitors, Earls of Chester, had always exercised the right to seize as forfeited any corn carried out of the city to be ground at other mills than the Dee Mills, in accordance with ancient custom. The whole of the defendants, with the exception of Widow Bavand, appear to have abstained from entering a defence, allowing judgment to go by default, or, as is not unlikely, arranging for a withdrawal; two of them, Dandrey and Ball, in fact, appearing in evidence against her. Margaret Bavand pleaded, after the usual fashion, that "such a prescription in favour of Dee Mills is injurious and against the law; for that a

III. DEE MILLS, CHESTER.

13. Action against the Abbey Mills, 1567.

prescriptive right cannot be in a matter of wrong." Evidence was called to prove:—

(1) That the multure and grist of the inhabitants of the city belonged and doth belong to Dee Mills. William Whitehead, baker, aged 82 years or thereabouts, has known for threescore years the customs to be as stated by plaintiffs. George Whitefield, miller, aged 68 years or thereabouts, confirms: about twenty years past he was farmer of this Bache Mill and the windmill, and at that time he "would not be supposed to grind any grain of the inhabitants of the city." Robert Dandrey, of the Ryddings, esquier [one of the original defendants apparently], aged 60 years or thereabouts, confirms, having known the custom fully for thirty-six years. John Cotgreave, miller, confirms.

(2) That corn and malt of the inhabitants of the city, carried to any other mills to be ground, hath been used to be [seized and] taken to Dee Mills; and the horses or carts that carried the same were brought to the castle and impounded. Thomas Ball, of Christleton [apparently another of the original defendants], aged 60 years, testifies that he hath been at divers times troubled for fetching corn out of the city to Boughton Mill, and about twenty-seven years past his horse with corn upon him was taken from him by the millers of Dee Mills. George Whitefield appears again, and says that when he was farmer of Bache Mill his horse and corn were taken as aforesaid; and he knows that the horse and corn of one Bushell, who was farmer of Boughton Mill, was so taken. William Whitehead and John Cotgreave confirm.

(3) That such persons as carried corn away from the said Dee Mills to any other mills were bound in recognisance in the Exchequer of Chester not to commit the like offence again. Robert Dandrey,

13. Action against the Abbey Mills, 1567.

esq., testifies that, when he was the deputy Chamberlain of Chester, divers times upon complaint of Ralph Goodman, divers persons who did carry corn and malt out of the city to other mills were sent for, and were by order and recognisance restrained and bound in the Exchequer from the same. George Whitehead states that when his and Bushell's horses were taken by the Dee millers, as aforesaid, they were both bound in recognisance of £10 to the king's use that they should not thenceforth so carry any corn; and they could not have their said horses or corn after it was ground till they had entered into such recognisances. An order of the Court was issued:—

> Upon hearing of the said cause, it was the 20th day of September, 9 Elizabeth [1567], amongst other things ordered and decreed that the said defendant, her servants and assigns, shall henceforth, during the interest of the plaintiffs in the said [king's] lease made to the said Sir R. Brooke, cease the carrying or fetching of any corn or malt of any citizens, within the city inhabitant, unto any mills without the same city, there to be ground, from the said Dee Mills: And that she, her servants, or any other by her procurement during the said time no further carry any the said corn upon the pains and forfeitures thereunto appertaining by the ancient custom used heretofore for remedy for the like offence.

Notwithstanding this edict, the widow persisted in getting what custom she could from such inhabitants of the city as were favourable to her and her cause. Consequently, by a second Bill of Indictment lodged against her three years later (September 15, 1570), it appears she, "in contempt and breach of such order, still doth daily by her servants and assigns fetch and carry corn and malt out of the said city." Whereupon the Goodmans plead, "the said breach of the said order is so notorious and daily iterated that, unless due and condign punishment be provided and had, it is like to grow to an ill example and great encouragement to wilful and disobedient persons to commit the like, to the contumacious and unlawful

III. DEE MILLS, CHESTER.

13. Action against the Abbey Mills, 1567.

defrauding and breach of all the orders of the Honourable Court." Also "the said complainants be daily endamaged hundreds of their multure by reason thereof and example of the same; so that unless some speedy remedy be provided they shall not be able to pay their rents, due and payable for the same." Evidently Mrs. Bavand had plenty of friends among the inhabitants; and making allowance for possible exaggeration of damages, the complainants no doubt felt the loss of multure she caused to be withdrawn to her "free mills."

Harl. MSS., 2081. 253.

The next step is to be found in the "Examinations taken in the Queenes Exchequer in her Castle of Chester" four days later (September 19, 1570), when every one of the widow's millers, carriers, and other servants ("great and small," as in Dublin in 1309) were carried off and examined as to the felonious practices of their mistress.—John Marshall, miller, has known the widow Margaret Bavand as farmer of the mills in question for seven years past. She took toll for corn ground. "She and her maid doth come every day as occasion serveth, and doth measure the same toll [corn] with a dish, and baggeth it up and carryeth the same away." Jane, wife of Richard Skagvener, "three score and od yeares or thereabouts," has known the watermill and the windmill beyond the Northgate above fifty years. Widow Bavand ground corn there brought on horses from the city. Thomas Mayre, carrier (who carefully states that he is now the "late carrier"), says he used to go daily by her commandment with two horses to fetch all manner of corn and grain from any of the inhabitants. "When wynd wantyd at the wyndmyll he carryed the corne to the watermyllne, and when water was scante he carryed the corne to the wyndmyll." Marquess Hogges was also a late carrier, but had never carried corn since

Text, II. 262.

September 4, and knew nothing about the matter; being evidently a specially unwilling witness. Ellys Anderton, carrier, John Huntingdon, blacksmith, and others gave further condemnatory evidence against the defendant.

13. Action against the Abbey Mills, 1567.

Margaret's contempt of court being proved, an order was issued on March 24, 1571, against her, John Marshall, her miller, Thomas Meire, her carrier, and three women-servants:—

> It appeareth by sufficient proof that the said defendant hath broken the former order [1567] in contempt of the Court: Wherefore it is ordered by the Right Honourable Robert Earl of Leicester, with the advice of the Council of the same Court, that process of attachment be awarded against the said defendant, returnable the twelfth day of April next. And at her appearance, to be committed to ward in the Castle of Chester, there to be punished for the contempt aforesaid. And that she shall remain in ward until she shall enter into bonds with sufficient sureties from henceforth not to infringe or break the said order by any means; but to perform and accomplish the said order according to the tenour and effect of the same. And also shall consent to pay to the plaintiffs the sum of thirty-five shillings for their costs and charges in the suit.

Harl. MSS., 2081. 109.

After this we hear nothing more of the brave-hearted widow, whose desperate attempt to earn a living for herself and her family at the expense of Dee Mills proved so utter a futility.

14. Introduction of the Gamulls, 1583.

14. During the progress of the preceding affair Ralph Goodman, in 1570, had died, his interest passing to William Goodman, who had joined Thomas Goodman in the prosecution of the Bavand suit. William Goodman, a prominent citizen, was several times mayor of Chester, and died during his year of office, in 1579. His widow, Elizabeth, in 1583, became the second wife of Alderman Edmund Gamull, who, by this alliance, was led to take that official interest in the mills which eventuated in his purchasing them and inaugurating the reigns of two of the most famous of all the king's millers of Dee.

III. DEE MILLS, CHESTER.

15. Soke damaged by Ald. Hankey, 1585.

15. Edmund Gamull speedily began the task of protecting the trade of the mills, calmly commencing operations in the year of his marriage by entering action against his old colleague, Alderman John Hankey, ex-mayor of Chester. This, his first campaign, was not fortunate. Public allegations had once more been made in Chester that the Dee millers were practising extortion; upon which John Hankey —who "used to be lessee of Dee Mills himself" as a marginal note on a brief informs us, though we find no record of the fact—had set up a horse-mill for the welfare of the community, and ground for the citizens at "the proper rate of $\frac{1}{16}$th." The same patriotic city father had already given evidence of public spirit in co-operating in 1567 with the mayor for the suppression of "the great strike" of the city bakers which had for a time threatened to starve out the entire city; and now his personal and practical opposition was entered against an alleged combination of compulsion and extortion at Dee Mills. Gamull took action against him in the Exchequer Court at Chester:—

Harl. MSS., 2081. 199.

"25 Elizabeth [1583]. Mr. Gamull exhibited a bill in Sccio Cestr. against one Mr. Hankey, an alderman of Chester, for erecting a horse-milne and withdrawing his owne and others' grist from Dee Milnes." In response to this, on 19 March, 25 Elizabeth, Hankey petitioned that "the defence of this suite might bee made a cytty cause [at the expense of the city], which the cytty denied, yet gave libtie. to his counsell to peruse the records of the cytty." "Whereas, &c., John Hankey has exhibited his Bill of Complaint in the Assembly against Edmund Gamull in 25 Elizabeth [1583], stating that he, being a free citizen, built a horse-mill on ground of which he had the fee-farme within the city; but that Mr. Gamull hath disturbed him of the use thereof, and will not

Ibid., 2081. 93.

Ibid., 2083. 602.

suffer him to have the multure of the said mill, but hath called him in suit in the Exchequer of Chester (and there in suit doth remain), contending that none ought to have any multure in this city but he, and thereby endeavouring to restrain other citizens as well as John Hankey; the latter praying it may be made a city cause: It was ordered that Mr. Hankey and his counsel, with Mr. Gamull, should be called before the mayor, and some means devised for peacefully settling the matter; Hankey to have access to the Book of the Inrollments of the Charters of the city in the meantime."

15. Soke damaged by Ald. Hankey, 1585.

The negotiation fell through; and it is to be gathered elsewhere, from an isolated order endorsed with the name "Hankey," that the usual discontinuance order of the Court of Exchequer was duly issued. But Hankey had clearly made out a strong case of extortion against the soke-millers; and they must have sadly suffered in reputation when, to this order, dated September 17, 1585, was appended by the Court a rider revealing strong sympathy with the defendant, who certainly would be ranked by the citizens as a considerable public benefactor:—

Harl. MSS., 2081. 199.

And it is further ordered by this Court that the said plaintiff [Gamull] nor any other occupier of the said Dee Mills shall not receive or take for the grinding of the said defendant's corn and grain or any other inhabitants of the said city any other greater toll than the ancient accustomed toll: Nor shall not extort or exact from the said defendant or any other inhabitants of the said city which shall grind their corn and grain at the said mills any other sum or sums of money, grist, or reward for grinding the said corn over and above the said ancient toll. Yet nevertheless, as the defendant hath been at costs and charges in the erecting of the said horse-mill and the provisions [plant] of the same: It is ordered by this Court that the matter concerning the same charges shall be remitted to the order of Robert Brooke, gent., and he to consider of the same; either that the said mill shall remain during the defendant's life, paying to the plaintiff and other fermors of Dee Mills a yearly rent for the same; or else the plaintiff to satisfy the defendant in money by way of sale of the said new-erected mill; or otherwise as shall be thought meet in equytye.

III. DEE MILLS, CHESTER.

15. Soke damaged by Ald. Hankey, 1585.

With this decision the Dee soke was practically broken. The mere contemplation of buying out the builder of an illegal mill, or compensating him for his expenses in resisting ancient custom, proved, indeed, that little rift in the lute which foreboded destruction. There seems to be no record as to what arrangement was made with Hankey; but so far as the citizens were concerned, he had gained a great victory for them at considerable risk to himself. As to Gamull, it is stated that he, in nowise discouraged, "afterwards erected a new mill adjoining to the said mills," and prepared for an increased custom from the city. In 1586 he became mayor of Chester.

16. Gamull's Purchase, 1600.

16. Convinced of the value of the establishment if the soke were duly enforced, Gamull, in 1600, became proprietor of the mills. Two or three different versions of the transfer are extant. In pleadings of which the date (by an allusion to a certain event taking place "about fifteen years ago") is to be fixed at 1635, it is stated that "in 42 Elizabeth [1600] Sir R^{d} Cotton's heire conveies the Dee Milnes under the yearly fee-farme rent of c^{li} to Thomas Gamull and his heires." Pleadings of about 1623 state, "George Cotton was seised of Dee Mills: the said George Cotton and R^{d} Cotton, his son, about the last day of June, 42 Elizabeth [1600], sold them to Thomas Gamull, gent., for £600, with the fisheries, which yielded £100 per annum"; Thomas, the son of Edmund Gamull, being at this date twenty-nine years of age. But in Edmund Gamull's own pleadings in 1609 it is said "the corn mills are held of his Majesty by knight's service in capite by Edmund Gamull, alderman, who purchased them for 3000^{li} in fee-farme of Mr. Cotton at 100^{li} rent per annum"; and at an inquisition held in the same year the jury find that the mills "are in the possession of Edmund Gamull, alderman, which,

Harl. MSS., 2081. 168.

Ibid., 2081. 154.

Ibid., 2082. 20.

Ibid., 2081. 216.

16. Gamull's Purchase, 1600.

with certain fishings, are held of his Majesty by knight service, and by the yearly fee-farme of c^{li}, payable to George Cotton and Richard Cotton, esqs., and their heirs." The tenure by knight's service—that of the previous holding by Sir R. Cotton—not only entailed upon the alderman a responsibility for possible military services, but had a more immediate effect in placing his grandson and eventual heir, Francis Gamull, in ward to the king.

17. Mills and Waterworks, 1601.

17. One of the most curious and disastrous attempts of Edmund Gamull to increase the soke of the mills was an alliance he made with the speculative John Tyrer, owner of city waterworks built closely adjoining the establishment of the king's miller. Gamull became a partner in this concern, agreeing to supply water or power from above the causeway in the river; while Tyrer, as was alleged, agreed on his part to supply no water to any citizens who refused to grind at Dee Mills. This unwise speculation was entered into in 1601; and, in the light of subsequent events, it is significant to note that in the same year some mysterious damage to the causeway threw Gamull's corn mills into idleness for several weeks. The full results of the league with Tyrer were not experienced by Edmund Gamull, as they accrued only after his death; but its early effects caused him the gravest anxiety of his entire milling career. The water supplying alike mills and waterworks being impounded by the large stone dam across the river, it was obvious that the abolition of this causeway would set the citizens free of both Gamull and Tyrer at one blow. The barrier across the river was in truth a structure easily open to adverse criticism. Nothing but the fact of its erection dating back prior to the strict enactment of 1352 (ordering the de- Text, II. 185. struction of causeways in navigable rivers erected

III. DEE MILLS, CHESTER.

17. Mills and Waterworks, 1601. after the reign of Edward I.) had permitted its existence for so long. Very much as we may see it at the present day sweeping across the stream, it effectually blocked all navigation of the Dee except for the shallowest of boats; and it was on this ostensible score that a cry for its abolition was raised. As, however, it existed perfectly legally under the Act of 1352, hostilities began with the assertion that, contrary to that Act, Gamull had raised it to a height greater than it originally had. Though no evidence

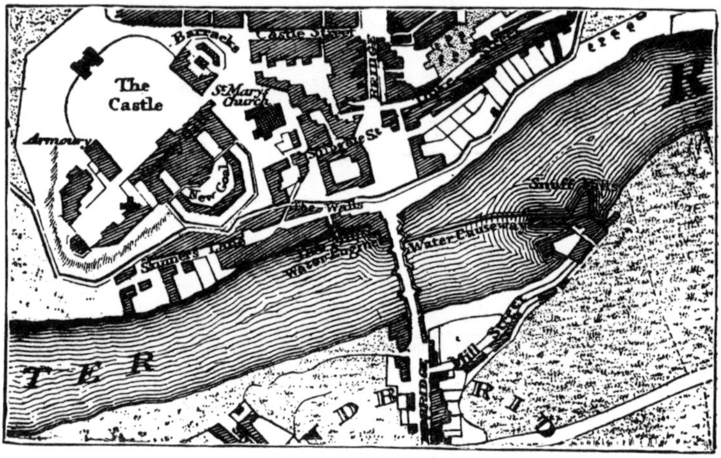

The Causeway across the Dee.

of this contention seems to have been produced, it was fostered not only by some citizens, but by certain of the neighbouring gentry, who, not entirely for public reasons, were resolved to ruin the mills whose soke the people could not break.

18. A Severe Winter, 1607.

18. On January 1, 1607, Tyrer, to discount the clamour, obtained the depositions of several of the oldest of the citizens on the matter:—

Harl. MSS., 2084. 178.

We whose names are subscribed at the request of John Tyrer have taken a perfect view of the causey at Chester and of the water that runneth over the same, which by reason of a very great frost long continuing is at this instant as low as it was this many

III. DEE MILLS, CHESTER.

18. A Severe Winter, 1607.

years at this time of the year: And now doth run over all the middle part of the same causey, where of late the breadth was in great abundance, but from that part upward towards the fulling mills the causey doth lie bare and without any water running over, because it is and hath been ever in our memories the higher part of the causey: Neither can any of us ever think or remember that any part of the said causey hath been raised or enhanced either by Mr. Gamull or any of the Goodmans or any other that were owners thereof, we being all of us near neighbours, and have known the same some of us this sixty years and all of us above forty years. All which we will be ready to testify upon our oaths as any occasion shall require. As witnesseth our hands subscribed this first day of January 1607.

Richard Hassenhall, of Handbridge, aged 82 years.
Christopher Pemberton, of the same, aged 64 years.
Thos. Ffletcher, glover, dwelling on the bridge, aged 60 years.
Otes Conelowe, beer brewer, aged 60 years or thereabouts.
Thos. Dannatt, glover, aged 60 years or thereabouts.
Edwd. Gill, glover, aged 80 years or thereabouts.
Edwd. Dalbye, glasior, aged 64 years.

At this juncture, also, the severe frost of the winter of 1606-7—"a greate frost that continued from Michaelmas to the middest of February"—was adroitly enlisted in his cause by the astute Gamull. For miles round the small rural mills were frozen into silence; Dee Mills the while, as though with supreme indifference to the icy grip of winter, were grinding grist busily and noisily as ever for all comers from the frost-bound city and countryside. A more valuable testimony to the public benefits of the king's mills it were impossible to produce; and Gamull did not fail to record it by depositions on behalf of his much-maligned establishment:—

Harl. MSS., 2083. 600[b].

Memo. That upon the xij daye of January they whose names bee subscribed, having beene upon occasion at Wrixham, in theyr returne they came by Sir Richard Trevor his milles at Morford, being in the highe waie side and aboute iiij miles from Chester: And because they saw the said milles standing still and not grynding, Richard Prymatt, one of theym who subscribes, tooke occason to questyon with the millner of the sayde milles and asked him the cause wherefore they wroughte not. Who answered him that all the saide milles (being iiij in number) had not wroughte for 5 or 6 daies then past; And the cause thereof was the froste. Then the said Richard Prymatt asked him wherefore they could not have broken and thawed the froste by some devise or meanes. Who

III. DEE MILLS, CHESTER.

18. A Severe Winter, 1607.

answered him it wold doe noe good, for they wanted water to serve theym withall. And then the saide Richard asked him howe they wold doe upon the like occason of want, yf Chester milles went downe: he answered he cold not tell howe they shold doe. But how longe after the saide milles stood still by the meanes above saied they cannot declare. All which they wilbe readie to depose as occason shall require.

RIC. PRIMATT.
JOHN WILDINGE.

Harl. MSS., 2083. 600^{b}.

Mem. Willm̄ Hunt of Chester having occason to travell into the countrie within vij or tenne miles of the citie did take vewe of all the water milles within that compasse, all of which stoode and were not able to grynde, by reason of the froste, and so had stoode by the space of a fortnighte at the least, not havinge water sufficient to serve them withall.

WILLM HUNT.

Decimo quarto die Jany. 1607.

Ibid., 2084. 176.

Richard Bridge, millwright, *and farmer and keeper of a certain mill at Stamford within the county of Chester, being late the inheritance of Thomas Venables, esq., deceased,** saith that he [Venables] did new build about six years since the said mills at Stamford. And that seven weeks since the frost that now remaineth did begin: and that the said mills would not go, by reason of the said frost, since the nativity of our Lord God [last past]. And that other mills also have stood for want of water: one at Hockenhall, two at Stapleford, one at Duddon, two at Barrow, two at Tervyn; which all stood for this month last past. He had taken corn to be ground to divers of the said mills, but could not get it ground, and therefore came to the said city the day and year above written and procured his said corn to be ground. And he saieth also that Mawde, the wife of John Dale of Stamford aforesaid, came within these four days last past to grind wheat here upon the like extremitie.

Others also came to these mills this morning—Mr. Smethwick of Duddon, Wm. Pulford of Cotton.

All standing idle: [a list of a score local mills].

Rob. Holkey's wife, of Grefford, came to Dee milnes on Fridaye ye xvij of Jan., and did paye for corne and malte, viz. 7 bʒ of malte and 3 bʒ of corne.

19. Destruction of Causeway ordered, 1608.

Text, II. 186.

19. More powerful influences than the foregoing, however, were directed to the ruin of the mills. The ancient jurisdiction of the Lord High Admiral over navigable rivers was now vested in Commissioners of Sewers; the local boards of which having authority over the Dee being those of Chester, Denbigh, and Wrexham. These declared their opposition to the

* The words here italicised are struck out in the original.

III. DEE MILLS, CHESTER.

19. Destruction of Causeway ordered, 1608.

continuance of the causeway. Gamull declared them to be actuated by purely personal motives; averring that certain of the commissioners were his own neighbours, interested in the closing of Dee Mills. As he subsequently pleaded, "Some gentlemen of the parts of Wales, having mills within four or five miles distant of the city, which they are desirous of improving, and having grounds upon the river Dee, do now, after 600 years, earnestly labour to destroy these mills and causey, being thus ancient many hundreds of years before even any Statutes of Sewers was made, and never since called in question." This assertion was certainly to an extent supported by a list of the local commissioners which he caused to be drawn up, showing that out of a total of fifty-five members no fewer than thirty-five were owners of mills or lands closely adjoining the Dee.* Foremost among them was Sir Richard Trevor, owner of Morford Mill which only recently Primatt and Wilding had sworn was frozen up; a certain petition by Gamull being endorsed, "Answer to one other petition lately preferred to their lordships by Sir Rd. Trevor, Knt., and others"; and a petition by the corporation of Chester, shortly to be mentioned, again singling out Trevor as the leader in the hostile movement. Gamull seems, therefore, to have had good reason to complain that "the commissioners are none other than the persons who had formerly taken private action against the mills for their own personal reasons; and they who are parties interested are also the judges." However, at Denbigh, the commissioners duly held their first inquisition on, apparently, (for in the various MSS. the dates seem somewhat confused,) August 29, 1608; and the jury found that the causeway was a hindrance to

Harl. MSS., 2082. 20.

Ibid., 2084. 234.

Ibid., 2084. 89.

Ibid., 2003. 37, 47.

* Another list of the commissioners, with a panel of jurors, occurs in Harl. MSS., 2003. 37.

III. DEE MILLS, CHESTER.

19. Destruction of Causeway ordered, 1608.

Harl. MSS., 2084. 183.

navigation and the cause of floods doing damage to the extent of £100 per annum. This was quickly followed by a second inquisition at Wrexham; when, a similar verdict being returned, the commissioners, "or some of them," made a formal decree that a break ten yards wide should be cut through the said causeway. This simply meant its destruction and the ruin of the mills; and against the order Gamull appealed.

20. Public Benefits, 1608.

20. An inquisition was held by order of the Exchequer at Chester, December 16, 1608, to enquire into the affairs of the mills and the royal local estate generally. During the enquiry, which was of a considerably detailed character, the "customs of Dee Mills" (as already quoted in full) were recited, and, after various evidences, the jury presented a verdict against the order of the commissioners being carried out:—

Harl. MSS., 2081. 216.

They say there are standing upon the north side of the said river, under the walls of the said city of Chester, five corn mills and one malt mill under two several roofs, in the possession of Edmund Gamull, alderman, which, with certain fishings, are held of his Majesty by knight service in capite and by the yearly fee-farme of c^{li}, payable to George Cotton and Richard Cotton, esqs., and their heirs. By which said mills so much of the said river is stopped as is sufficient for the maintenance of the said mills with water. The inhabitants of the said city do for the most part grind their corn and malt at the said mills with great ease to themselves, especially the poorer sort, who usually grind their pecks and small portions of corn without paying any toll for the same. And also the Company of Bakers within the said city, being in number twenty householders and above, that are sworn and bound to serve the city with sufficient and wholesome bread, have their corn ground at the said mills every market day after the said market is ended, so as the poor men having small stock may make short returns thereof weekly and maintain themselves and their families, and serve the city with bread for their necessary sustenance. And divers also that dwell in the hundred of Wirrall and other places within the county of Chester, six, seven, and ten miles distant from the said city, do often grind their corn at the said mills, and have been heretofore accustomed so to do; by reason they have very few watermills in Wirrall. Within which time also great quantities of corn oftentimes during the late wars in Ireland have been ground at the said mills and sent over in meal into Ireland for the supply of the armies and garrisons there.

Text, ante, p. 76.

20. Public Benefits, 1608.

And that the owner of the said mills receiveth by the same a good yearly profit, and maintaineth many millers, carriers, carpenters, and other persons in work. And that there are no other water corn mills or windmills within the liberties of the said city, but only [the Abbey Mills, viz.] one watermill which standeth upon a little brook, called the Bach Mill, which often wanteth water, and one wyndemille which often wanteth wynde. And they say also that about three years since, during the late visitation of the plague, which continued in the said city three years together, the inhabitants within the said city were restrained from grinding their corn at other mills in the county: so that if the said mills of Dee had not been, they should have been (as they are persuaded) in great danger to have wanted bread: they also of the country forbearing to come into the city as they had formerly done before the sickness.*

They further say that in part of the stream of the said corn mills there was lately built by John Tyrer two wheels with certaine engynes for the conveyance of water into the said city.

Lastly, they say that the said mills were anciently parcel of the possessions of the earldom of Chester, and divers of the said mills have continued time out of mind and were standing, as appears by divers records in the time of Richard, Earl of Chester, in the year of our Lord God 1119. And also in the time of Randle, Earl of Chester. And that in or late before the time of John, Earl of Chester, called John de Scotia, the number of the said corn mills were increased by the building of new mills when the said Earl granted the tithes to the abbot of Chester. And that it doth not appear unto them by any proof or otherwise that any of the said corn mills now standing were built since that time; but how long before or when they were built it doth not appear. Saving that about vij or viij years since the said Edmund Gamull did build one of the said corn mills. By reason of which mill or by the other two waterwheels before mentioned [the waterworks wheels] we do not find the causey to be enhanced or the passage of the water to be hindered more than before.

And that the Dean and Chapter do now receive £5 per annum in respect of the tithes of the said mills and fishings; and that the total profit per annum received by all the mills and the fishings, causey and waterworks, is £1000.†

* "1603. The 22nd August in the night time a wonderful exhalation of a fiery colour, likewise a canopy, was seen over this city, and in September following the great plague began in Chester. Infected persons were taken out of their houses, and conveyed into houses and cabins at the waterside, near unto the New Tower.

"1604. The plague was very hot in Chester, and a great many were sent into cabins, and a great number of citizens removed into the country. There were no fairs kept, neither did the watch go on Midsummer Eve.

"1605. The sickness of plague continued till about the middle of this year. The total that died of the plague only is about 1313 persons."—*Annals of Chester.*

† Regarding the fulling mills across the bridge, the jury state that "they keep very many in work, to the great furtherance of the trade of clothing in these parts . . . there being no other fulling mills that they know of in or near Chester."

III. DEE MILLS, CHESTER.

20. Public Benefits, 1608.

In view of the forthcoming appeal to the Privy Council, Sir John Sauvage, mayor, with the aldermen and commonalty, presented also a breviat and petition to the King, the Privy Council, and the Lord Chancellor, praying that the Commissioners of Sewers might be superseded, and their decree for the destruction of the causeway nullified. The petition was directed "particularly against Sir R. Trevor and other Welsh gentlemen who fancy themselves damnified." It was shown that the verdicts of the inquisitions at Denbigh and Wrexham were at variance with that held at Chester; that persons who had originally been interested to solicit opposition to the causey had since been enabled to prosecute it; that out of fifty-six commissioners only eighteen of Wales and six of Chester had agreed to the decree, and many of these had not been present at the meeting, but their hands and seals had been obtained afterwards; while the commissioners had decreed on matters impertinent to their charge, tending to maintain the private good of some citizens and not the public weal and benefit of the city. The commonalty, it was urged, had many reasons for wishing the causeway preserved: the haven was in but the same condition as described by Geraldus Cambrensis (in the 12th century), while his Majesty's customs at Chester were greater than ever: the causeway, moreover, had been erected six hundred years, and was never complained of before, and could not be abated without destruction of the whole of it and loss to the mills.

Harl. MSS., 2003. 37, 47.

The chartered Company of Bakers were equally zealous in protesting against the decree of the commissioners. A petition by them set forth how they who ground there were sworn to serve the city with good and sufficient bread; how the ships of the king and of merchants were supplied from these mills; how

Ibid., 2084. 345.

20. Public Benefits, 1608.

they themselves maintained their households (numbering in all one hundred persons*) by trade depending on the mills; and how this would be otherwise impossible for them, since every market day they bought corn and had the sole use of the mills for grinding the same: moreover, the mills yielded a large rent to the king, and were heavily chargeable for subsidies; while, further, they ground corn for the king's ships and for the soldiers in Ireland.

Meanwhile, the agitators for the destruction of the causeway became impatient at the delay, and threatened "to come very speedily to Chester with great multitudes to put the decree into execution; so that the memorye of the antient enmitie and hostilitye between the English and the Welsh is revived."

21. The Case, 1608.

21. In preparation for the appeal, Gamull fortified himself with the opinion of no fewer than thirteen of the foremost legal lights of the day, all of whom expressed a general decision in favour of the legality of the causeway. These documents—each with the opinion briefly scribbled at foot, and one bearing the signature "Fr. Bacon"—comprise the merits of Gamull's case, which was thus stated:— Harl. MSS., 2084. 237, 319.

* The names of the Companie of the Bakers, with theare servauntes and houshoulde people. [*c.* 1650.] Ibid., 2054. 44

Hughe Crumpe his househould people are	x	John Maddocke	vij
Richard Wryght his people ...	xiiij	John Blanchard	viij
William Edwards	viij	Robt. Couper	viij
William Bennet	viij	Widd. ffearnall	vj
William Kinge, Senior	x	Roger Kinge, Junior	xiij
William Maudstey	ix	Widd. Kinge	iiij
Thurstane Holde	vij	George Blynston	v
William Wryght	ij	Raffe Mosse	viij
Widdowe Adams	vj	John Couper	vj
John Welshman	vj	Richard Hall	iij
Raffe Ctie	iiij	William Kinge, Junior	viij
John Rigmayeden	viij	Robt. Lee	v
Widdow Cross	iiij	Roger Whithead	vij
Roger Kinge, Senior	xiiij	Widd. Johnes	iiij
John Ashton	xvij	Widd. Carter	v
Eduard Blynston	iiij	Widd. Coup[er]	iij
		Widd. Deanson	iij

These are the whole houshould people of bakers.

[Total, 33 bakers, with 234 dependants.]

III. DEE MILLS, CHESTER.

21. The Case, 1608.

(1) The statute of Magna Charta, made Anno 9 Henry III., is quod omnes kidellj deponantur de cetero nisi per costa maris, &c. [that all kidells or causeways be taken down, &c., except those on the margin of the sea, &c.].

(2) The statute of 25 Edward III., cap. 4, is that all mills, weirs, and causeys set up during the time of Edward I. shall be pulled down.

(3) The statute of A° 1 Henry IV., cap. 12, recites and confirms the statute 25 Edward III., and appoints weirs and causeys built before the time of Edward I. to be surveyed by commissioners: and those that be too much enhanced since their erection to be amended to the old-time level.

(4) By the statute 23 Henry VIII., cap. 5, Commissioners of Sewers appointed to survey weirs and causeys; not making mention of any time when they were built: and those they find excessive and hurtful they are to pull down.

Questio. The question is whether a causey or mill of the inheritance of the King's Majesty, or any other which hath continued time out of mind, and may be proved by records to have been standing in the time of William Rufus long before any Statute for Sewers was made, and which hath not been since enhanced nor ever called in question, may now be lawfully pulled down by the Commissioners of Sewers: And if they offer to pull it down, what may the owners do justly by law?

Bacon was of opinion that "The Commission of Sewers and the statute of 23 H. 8, upon which it is grounded, extends not to this case." Sir John Bridgeman wrote: "The commissioners ought not to meddle with any such weir or causey: if any attempt to pull it down, the owner may resent them." Other learned counsel expressed similar views.

22. Commissioners' Decree quashed, 1609.

22. The cause at length came before the Privy Council, April 27, 1608, and was referred on points of law to three judges, including the famous Coke; who, upon May 29, 1609, decided that the commissioners had no power to issue their decree. The latter was thereupon quashed by an Order in Council, June 2, 1609:—

Harl. MSS., 2084. 336.

At the Court at Whytehall, 2 Junij 1609.

In the cause depending between the lords of his Majesty's most honourable Privy Council concerning a decree heretofore made by certain Commissioners of Sewers for a breach of ten yards in length to be made to the bottom of the river in a causey of stone within

22. Commissioners' Decree quashed, 1609.

the river of Dee and city of Chester serving for the maintenance of the fulling mills of his Majesty, and of divers corn mills and waterworks of others standing at the ends of the said causey: Against which decree the mayor and citizens of Chester and likewise his Majesty's fermors of the said fulling mills and divers others intested in the said corn mills and waterworks did make complaint unto their lordships by their several petitions setting forth the great benefit and commodity redounding to the said city and country adjoining, and to the said petitioners and many others by means of the said causey, mills, and waterworks: And further alleging that the said breach cannot be lawfully made in the said causey by that the same hath continued from time out of mind:

Whereas the said cause being heard and long debated before their lordships the 27th day of April, 1608, in presence of divers of the said parties and their counsel, it was then ordered by their lordships that, for their lordships' satisfaction on the point of law concerning the validity of the said decree and authority of the said commissioners, letters should be written to Sir Thomas ffleming, Knight, Lord Chief Justice of England; Sir Edward Coke, Knight, Lord Chief Justice of His Highness' Court of Common Pleas; and Sir Laurence Tanfield, Knight, Lord Chief Baron of His Highness' Exchequer, to hear the said cause and the counsel learned on both sides, and to consider the said decree of the commissioners and the laws and statutes whereupon the same is grounded, and to certify thereupon whether the said decree and proceeding of the commissioners be warranted by law:

According to which said order, letters being sent to the said Lords the Judges, they have heard the said cause and the counsel learned on both sides: And upon perusal and consideration of the said commissioners' said decree referred unto them, and of the said laws and statutes, they have lately certified their opinion, dated the 29th day of May last past: Which said certificate being received and this day weighed and considered:—

Forasmuch as it appeareth to their lordships that power of the said Commissioners of Sewers doth not extend unto the said causey, the same shall not be pulled down . . . the commissioners shall proceed no further against the causey and mills . . . and their said decision shall not be put in operation.

The report of the three judges, with their signatures attached, appears in another document:—

Harl. MSS., 2084. 363.

The law doth not extend to such ancient mills, millstanks [pools], and causeys as were erected before the time of King Edward I., except they have been enhanced and exalted above their former height, and thereby made more prejudicial and hurtful than formerly they were. In even which case they are not to be pulled down from the top to the bottom (as the said decree now questioned doth appoint), but to be reformed by taking away and abating the excess and enhancement only. And so we humbly take our leaves.

Serjeants' Inn, ffleet Street, 29 May, 1609.

III. DEE MILLS, CHESTER.

23. Francis Gamull, heir, 1616.

23. Edmund Gamull, with much ingenuity and harassing toil, had now placed the fortunes of the mills upon an apparently sound footing. Their ancient rights were fully confirmed, the Commissioners of Sewers were defeated, the citizens of Chester were held in friendly alliance, and the income was sufficiently high to yield an assessed rental of £1000 a year. Seven years of peaceful prosperity for Gamull ensued, broken only by the loss of his eldest son, Thomas Gamull, Recorder of Chester, who died in 1614. In the church of St. Mary's-on-the-Hill, near the mills, was erected to his memory "a fair and beautiful tomb,

very curiously and all of alabaster wrought." Here two years later the alderman himself was laid; leaving as his heir his grandson Francis (son of Thomas, and represented on the tomb as a child kneeling at the feet of his parents), subsequently the last and greatest of the feudal millers of Chester. Of Edmund Gamull—"a late alderman, of great and good account, of this city, . . . for whom were to be wished some monument answerable to his worth"—there is no monument.* Though the effigy of Alice Gamull, widow of Thomas, (who erected the tomb,) is shown thereon, she survived her husband some years, and, by a second marriage, introduced

* In the same church, we may not forget, are buried the two Randle Holmes, justices of the peace and mayors of Chester; to whose praiseworthy care in preserving the scattered milling documents of the Gamulls, now contained in the Harleian MSS., this sketch of the history of the mills is so greatly indebted.

another manager to her son's milling estate. This was Edward Whitby, who had succeeded her late husband as Recorder of Chester, and who, in due course, was appointed trustee for the young heir, Francis. The latter, as already mentioned, was a ward of the king in virtue of the tenure of the holding of the mill; and it was probably this circumstance that in after years rendered Francis Gamull a generous and noble supporter of the losing cause of Charles I., at a cost for which, in itself, the knighthood bestowed upon him by the king proved but a very doubtful equivalent. During the minority of the heir all actions were taken in the name of his trustee, Whitby, as comptroller of the mills.*

III. DEE MILLS, CHESTER.

23. Francis Gamull, heir, 1616.

24. The new comptroller fully kept up the Gamull traditions. In the winter of 1622–23 he issued a process in the Exchequer Court against several small tradesmen of Chester, who, doubtless from motives of economy—for they admitted themselves to be "of the poorer ranke"—ground at some cheaper mills. The defendants, in abject helplessness, twice appealed to the City Assembly:—

24. The Poor Tradesmen, 1622.

30 Jany. 1623. The petition of Nicholas Radcliff, and Mary, his wife; Raphe Wilson, tailor; Thomas Johnson, cutler; and Raphe Davis, all free citizens of Chester, to the mayor and aldermen of Chester, sheweth. They have been served with process out of his Majesty's Court of Warde and Livery at West^{r},† at the instance, they suppose, of Mr. Edward Whitbye, guardian of Francis Gamul, gent.,

Harl. MSS., 2081. 171.

---

* Annexed is a similar instance of trusteeship for royal wards, heirs to milling property:—6 October, 13 Elizabeth [1571]. At the motion of Mr. Barton (trustee), for and in behalf of Anne Riggley and Katherine Riggley, the Queen's Majesty's wards, it is ordered that an injunction shall be directed to Robert Mawson (and others) of Riggley, in the county of York, commanding them thereby, from and after the receipt thereof, to grind their corn and grain at Riggley Mills, parcel of the said wards' inheritance, in such sort as they did grind their corn there at the time of the death of Henry Riggley, Esq., father of the said wards.

Ibid., 2083. 525.

† This court was instituted by statute 32 Henry VIII., ch. 46, to superintend and regulate enquiries upon the death of any of the king's tenants as to heirship, &c., and to overlook the control of the estates of the king's wards. It was abolished soon after the Restoration.

III. DEE MILLS, CHESTER.

24. The Poor Tradesmen, 1622.

his Majesty's ward, for their appearance, &c. They pray that, as they are free citizens and tradesmen, they may not be put to unnecessary expense by proceedings at law: that the mayor will move the Recorder to stay proceedings against them; and what can or may bee justlie alledged agaynst them they will referre themselves to y[r] Worshipps mature and deliberate censures. They suppose they have power to grind their own corn for their own use or benefitt at anie milne for theire best profitt, ease, and advantage: and desire the advice and assistance of the Assembly, by which they will be bound, as they are of the poorer ranke, and not able to stande out in soe high a court and uppon such greate charges as this suite will requier; while their cause is the cause of all the citizens.

The sequel does not appear, but it needs no conjecture; as, a dozen years later, copies of the petition were made by the Dee millers from the entries in the City Assembly Book, and produced in support of their own case against Russell and others.

25. Troubles of a Sub-lessee, *c.* 1622.

25. To this period is assignable an undated paper containing "Mr. John Brerewood's allegations that Dee Mills should pay nothing to the repairs of St. Mary's Church," which is situated on the castle rock just above the mills. John Brerewood (son of Robert Brerewood, glover, thrice mayor of Chester, who died in 1600*) seems to have rented "one of the six corn mills" about 1622; and in justification of his protest against the imposition of a church rate drew up a kind of income tax schedule of his own devising:—

Harl. MSS., 2082. 6.

First, by the deeds of the purchase [E. Gamull's] the mills are not mentioned to be in any parish, but only standing on the water of the Dee. And if they be in any parish they must be in the parish of St. Werburgh [the cathedral], because it doth appear by several records that in anno 1119 Richard, then Earl of Chester, granted a rent to the then abbot for the tithes, which rent is now received by the Dean and Chapter. This was long before St. Mary's was built. Secondly, the profits of Dee Mills ought not to be assessed: because it is a personal thing, and the profits do arise by great charge and expenses of the owners, who, besides their charge, pay 105[li] rent per annum. And a mylner is an Art and facultie and an Handicraft within the statute of 2 Edward VI., cap. 13. See my lord Cookes 2nd Instit. in his comts. upon *Articuli cleri*, cap. 5, page 621. Thirdly, [receiving a rent out of a parish does not make a man a

* "1600. Mr. Robert Brerewood, mayor, caused the bakers to bake farthing cakes; wheat being at 12s. the old bushell."—*Annals of Chester*.

parishioner]. Fourthly, for my part I have only one of the six corn mills, for which I pay 21^{li} rent per annum, and I keep an Horse, which doth cost me in his keeping at least 7^{li} per annum. And also I keep two men and a woman that sells my corn, which receive in yearly wages from me at least 22^{li} per annum, besides milstones, candles, sope, tallow, and repaires of the mill, which I can safely make Oath of, is more then I get by my mill.

26. Action against Boughton Mill, 1623.

26. Scarcely were the foregoing affairs cleared up, when, in 1623, the widow of William Davenport, gent., was found to be grinding malt for citizens at her rural horse-mill at Boughton. This had been built by her late husband in 1620, and had been constantly worked by him. Upon his widow continuing the trade, however, she was served with process, and a "Writ of Commission out of the Prince his highness Court of Wardes and Livery" was promptly directed to four justices to examine her upon the charge.

Harl. MSS., 2081. 120.

This preliminary enquiry took place at the house of William Fisher, innholder, September 22, 1623; but Jane did not state very much. She was forty-four years old, and knew several mills (which she named) in and near the city that had been working for years; one for at least twenty-five years: corn had been carried to the Boughton mill without interruption away from the Dee millers, and till this present affair she never knew of any such complaint or interruption from them: she had never heard of any one being fined for not grinding at Dee Mills: the rate of toll at her mill she did not know, but believed her servant there took one measure for twenty and four*: her mill was more profitable than the Dee Mills to the maltsters in grinding malt by reason of the closeness thereof: she had never used any words of discommendation concerning either the grist or the taking of toll at Dee Mills. The inference from these statements appears to be that the widow had been allowed to work the mill until her tongue had run away with her

* The rate at Dee Mills being one in sixteen.

III. DEE MILLS, CHESTER.

26. Action against Boughton Mill, 1623.

discretion, and her insinuations against the powerful soke mills had ensured her downfall. If so, Recorder Whitby could scarcely avoid taking action, and we may understand his "displeasure," which the widow said she could not.

The case was duly opened November 8, 1623, and Mrs. Davenport proceeded by petition with her defence.

Harl. MSS., 2081. 120.

She severally sayeth that she verily thinketh that the said information is exhibited against her by the only means of Edward Whitbye, esq., upon some displeasure by him conceived against her of purpose to vex and trouble her, and to put her to unnecessary costs and charges without any good or sufficient cause at all to do so. Which she the rather thinketh, because there are many others named defendant in the information, yet none of them, to her understanding, are served with any process to answer the same; but only this defendant, who by reason of the weak estate she was left in by her late husband is least able to defend the suit. . . . She further sayeth that William Davonport, gent., her late husband, about three years last past for his own better advantage did erect this one horse-mill for malt upon a parcel of land his inheritance, and made the same over to her in jointure on their marriage. Which mill he did quietly hold and enjoy while he lived, receiving the multure and causing to be carried and recarried malt, without disturbance. And about two years past he died, and this defendant was content to sell the better part of her jointure to pay his debts in order to keep the mill for the support of herself and her children. She had caused to be carried such malt as any of the citizens or others would grind at her mill without disturbance until late: when as she verily believeth the said Edward Whitbye, esq., upon some causeless

26. Action against Boughton Mill, 1623.

displeasure and wish to oppress and impoverish her, caused this information to be exhibited against her—who is the meanest and least able to defend this suit of all the others that have mills in or near the said city. Wherein she humbly prays the favourable consideration of this Honourable Court.

In her pleadings she formally entered a denial of the exclusive rights of Dee Mills, in the stereotyped manner which more or less appears with constant reiteration through all the defences entered against the Gamull suits:—

This defendant will aver and prove that there now are and, by all the time whereof the memory of man is not to the contrary, have been divers and sundry mills, as well watermills, as windmills, for the grinding of corn and malt within the said city and suburbs and liberties thereof, other than the aforesaid Dee Mills. And that also there now are and have been, for and by all the time of man's memory, divers and sundry other watermills and windmills for the grinding of corn and malt, of the inheritance of sundry persons within four or five miles distant of the said city of Chester. And that during all the time aforesaid not only the freemen and citizens of the said city but the inhabitants and dwellers in the same and the liberties and suburbs thereof have at their wills and pleasures ground and sent their corn and malt to be ground at the said other mills as they best liked or pleased. And the owners of the said other mills, and their servants and tenants and their carriers and servants, by all the time aforesaid have used and been accustomed respectively to fetch and carry into and out of the said city and the suburbs and liberties thereof at their pleasure all and every the corn and malt of the aforesaid citizens or other persons as would or did grind at such then said mills respectively without any lawful gainsaying of any of the owners, farmers, or occupiers of the said Dee Mills.

There had, however, been "lawful gainsaying" of a very decided kind against such practices; and there can be no doubt that the usual order of the Court was made, and that Jane Davenport's mill went the way of that of Margery Bavand.

27. Waterworks Dispute, 1634.

27. Francis Gamull, now of age, became mayor of Chester in 1634. Whitby remained associated with him in the management of the mills, but the first

III. DEE MILLS, CHESTER.

27. Waterworks Dispute, 1634.

Harl. MSS., 2083. 450.

Ibid., 2082. 34[a].

contest in which they conjointly engaged resulted in a defeat that still further disintegrated the compulsory soke. The alliance of the mill with Tyrer's waterworks had continued from the time of Edmund Gamull, and Tyrer faithfully carried out his part of the bargain in refusing water from his pumping-works to absentees from the mills. That he did so, occurs frequently in evidence; in the pleadings of one, Thomas Webster (1637), for instance, he states that "formerlie he in curtasie and upon threats to take away the water from him did grinde at Dee Milnes, where he receyved greate wrong and prejudice." Whitby must surely have seen that this attractive but illusory scheme of co-operation could only expose the frailty of the soke of the mill; but he seems to have preferred pinning the citizens to the mills at one operation by the agency of Tyrer to encountering them singly in costly law suits. The plan succeeded till the death of Tyrer in or about 1634, and the opening of negotiations with his son by Francis Gamull and Whitby for the purchase of the waterworks. This step may have been due to the injudicious management of Tyrer junior; as in 1634 Randle Holmes is found petitioning Gamull, then mayor, complaining that the proprietors of the waterworks were charging greater rents than before, and had wrongfully cut off the service-pipes. Others, however, had an eye to the purchase of the coercive milling factor; and during Gamull's absence from Chester the concern was acquired in the public interest by Sir R. Mainwaring, ex-alderman of Chester, and friends; the agreement stipulating that the purchasers should pay the owners of the mills a rental of £10 per annum on a lease of ten years. But Mainwaring and his partners at once cut the tie that bound the waterworks to the mill; "uttering and giving out speeches against the

Dee right of multure, saying it was a slavery and a bondage for men to be bound to grind at any mill; and that they were able to discharge the greater part of the citizens from the same." An action was thereupon entered against Philip Mainwaring, Robert Harvey, alderman, and Thomas Alderney, apparently for breach of agreement; the case being entered for hearing as usual before the Chamberlain of Chester. The interrogatories in the suit included:—"Is there less toll taken at these mills than is ordinarily taken at other mills? Is not the sixteenth taken for the toll of malt and corn? Did not Tyrer refuse water to any of the inhabitants who would not grind at Dee Mills?" The mills were described as "the said six mills called Dee mills . . . corn mills all under one roof . . . malt mill in building adjoining . . . waterworks in the court adjoining the malt mill." The matter was ultimately heard in the Star Chamber, May 6, 1635, it being held that the owners of the waterworks were entitled to receive their supply from the mill-pool on payment of the old rent; but the crux of the whole matter, whether they should force their customers to grind at Dee Mills, was left for decision by the Court of Exchequer. I have not been able to trace a record of this decision, but apparently it was adverse to Gamull, as no more is heard of the old standing alliance; and, indeed, its extinction was much to be desired.

27. Waterworks Dispute, 1634.

Harl. MSS., 2081. 38.

Ibid., 2081. 36

28. While the struggle with the owners of the waterworks was in progress, some far-seeing citizens, anticipating the destruction of the Dee soke, prepared for the expected free-trade in grinding by erecting horse-mills. Nothing daunted, Gamull took steps against one of them, Edward Russell, in 1635; presenting a petition in the usual way to the Chamberlain of Chester. The existing copy of this document, one evidently made for the defendant, Russell, contains

28. Another "Case," 1635

III. DEE MILLS, CHESTER.

28. Another "Case," 1635. Harl. MSS., 2081. I.

various marginal annotations by counsel on his behalf; these being here appended in footnotes to an abbreviation of the deed :—

To William, Earl of Derby, and James Stanley, Lord Strange, Chamberlain of the city of Chester. Your orator, Francis Gamull, son of Thomas Gamull. . . . Whereas our late sovereign lord, King Edward VI., &c., was seised of certain ancient water corn mills called Dee Mills . . . which mills are very fit and subservient for the inhabitants and residents of the said city and suburbs and liberties.* . . . All the above are obliged there to grind all grain and corn which they expend in the city . . . and have paid and ought to pay for the grinding a certayne toll to the tenants, farmers, and occupiers.† . . . No persons to have or use any other corn mills in the city except the Abbey Mills, known by the names of ye Windmiln withoute the Northgate and ye Watermylne called the Bach Mylne.‡ . . . No person ought to erect any new mill in the city or liberties, so that any grist, multure, or other benefit be withdrawn from Dee Mills. No persons ought to carry corn, expended in the city and liberties, without the city to be ground at other mills. And whereas King Edward VI. did by letters patent, 24 June, 7th of his reign, grant to Sir Richard Cotton and Dame Jane his wife all his said corn mills,§ with all rights and appurtenances and other profits and commodities ‖ . . . and all soke, suit, and services of the same.¶ . . . In 42 Q[n] Elizabeth the mills were acquired by Edmund Gamull.** . . . Since then Edward Russell and five others, inhabitants of Chester and bound to grind at Dee Mills, and W[m] Reibe and R[d] Sharples, who allege themselves to be farmers of a water corn mill near Chester,†† have conspired to grind at this said mill. . . . About a month ago Edward Russell hath built a new horsemill in Chester, and doth grind malt and oats for the inhabitants: and Reibe hath also ground during a year past great quantities of grain taken to his mill out of the city in carts and brought back again to the city after. . . . Other mills also have been lately erected near the city. . . . An order is prayed by your orator restraining them."

---

* "He hath but one malt mill, and that by no means can serve the town. Nor ever did."

† "His predecessor did favour one or two mills; till [when] he had purchased Dee Mills, he did demolish them and built at Dee Mills."

‡ "He himself builded one, and so have divers others and used them. There was always other milnes. As 2 windmills at Boughton. The one the Deans and Chapters: the other his predecessors took and used and demolished and paid for it, on purpose to get more custome to Dee Mylnes. There were within memory on or near Chester iiij windmills and vij horsemills."

§ "Not how many."

‖ "Not custom of multure."

¶ "No custom."

** "If it be customary to grind at his mylnes, then might he bring his action against them that owe it: but not agaynst mee for erecting a milne."

†† "Four miles distant."

28. Another "Case," 1635.

In due course "Mr. Francis Gamull exhibited his Bill in the Exchequer at Chester, to be reteined against the said Russell and others, who have set up mills and withdrawn their grist from Dee Mills: and upon view of the record of the custom and former decrees and affidavit of possession, some of the defendants were accordingly ordered to grind at Dee Mills till the hearing of the cause." Gamull thereupon submitted another case for counsel, choosing for the purpose William Nuthall; whose opinion is annexed:— Harl. MSS., 2081. 170.

1. Whether the tenure in capite, being aided by the said decrees, may preserve the ancient rights of multure, and doth so bind the inhabitants of the city to grind their corn and malt at Dee Mills, as the now owner may be reteined in equity in his suit, though the king have neither reteined rent nor tenements; nor other mills should be erected in prejudice thereof?—I conceive ye Court of Chester by reason of their former decrees have a full privilege to make the like decree againe. Ibid., 2081. 170.

2. Whether the sufferance of the said horsemills being erected within memory, and most of the city grinding at Dee Mills, hath destroyed the customs belonging to Dee Mills?—Y^e^ sufferance of ye horsemilne hath not destroyed ye custome.

3. Whether the addition of a new milne hath destroyed the custom belonging to the old?—It hath been many tymes resolved in ye Excheq^r^ at West^r^ that ye erection and addition of a new mill destroyed not ye custome of ye old, but ye tent^s^ are not bound to grind at ye new milne, but must bring their grist to ye old mill, and it is not much materiall where ye miller grindeth it

4. Whether the fermors having not of late kept courts nor seised corn according to the ancient customs have lost the same by non-user?—Y^e^ non-user doth not destroye ye custome.

5. Whether upon view of the former decrees and ancient customs and affidavits that the defendants did formerly grind at Dee Milnes, the Court may enjoin and order them to grind there till the hearing of the cause and other order?—I doe not conceave the Court before a hearing ought to require the tent^s^ to grind, for in that case it is ever sayd y^t^ ye Court should in effect determine ye cause before it is heard.

6. Some of the defendants who were ordered to grind at Dee Mills have, to evade the said order, ground their malt at the said horsemills and given out they buy their malt ready ground and not before. Soe what course is to be taken herein, for it was never known in these parts that any ever bought malt ready ground?—If the tenants be such as are bound by ye decree, an affidavit might be made of it, and then ye Court will get an attachment: uppon which attachment ye party is to be examined upon interrogat^ns^ if it appeare to ye Court this is a contempt, and as a contempt is punishable.

III. DEE MILLS, CHESTER.

28. Another "Case," 1635.

Harl. MSS., 2081. 168.

From the brief for Russell's defence it appears he pleaded the "ancient customs" of Dee Mills, upon which Gamull relied as to soke, to be not a charter upon record in the Exchequer, as alleged, but only a private document belonging to the mills "when they had a court": also, that there was no such custom in Queen Elizabeth's time; that "there are not so many mills now in and near Chester as there have been, for there were twelve, and now there are but six besides Dee Mills, and the latter are insufficient to serve the city, and they make bad usage; further, E. Gamull bought nothing but the mill structure, neither courts nor privileges, nor any warranty for the same: nor hath he kept any such courts since the purchase." Of the trial and the result there seems to be no available record, but Russell and the others were certainly ordered to suppress their mills.*

29. Buying Ground Malt, 1637.

29. The sixth of the interrogatories in the preceding case discovers in practice the same ingenious evasion of soke which had been prohibited at Liverpool in 1524, at York in 1617, and other places. The Chester maltsters about 1635 had combined to purchase outside the city malt ready ground; a process which they imagined might free them from the Dee soke.

A small ring of maltsters was formed; a subscription was raised to procure legal advice from London; John Edwards, a miller of Boughton, joined; and one, Tyrer, a malt-dealer, began openly to sell the ready-ground malt. On May 13, 1636, an order was secured against four of the ring—Thomas Dennison, Richard Dickenson, Edward Russell, and Andrew Minshall; but despite this, not only did they carry on their illicit practices as before, but they were joined by four others—Thomas Weaver, Henry Hazelwell, Elizabeth

Harl. MSS., 2081. 111.

Ibid., 2083. 450.

* The only apparent clue to the date of this affair is an allusion in the case to the erection of Davenport's mill in 1620 as being "about fifteen years ago."

III. DEE MILLS, CHESTER.

29. Buying Ground Malt, 1637.

Pemberton, and Katherine King; all of whom were thereupon included in another indictment. The case against Dennison, and his defence, are the most interesting of the series. The bill formulated against him stated that since the order of May 13, viz. between the 20th and 25th of May, he had ground two brewings of malt, containing forty-eight measures each, at Edwards' mill at Boughton. Against this practice it was pleaded that the buying of ground malt was fraudulent and a breach of the order of the Court of May 13. Richard Gough would testify that on May 24 he saw six of Dennison's sacks of malt at Edwards' mill, and witnessed them carried the same night to Dennison's house. William Culcheth would prove that he had several times tried to serve the order of the Court of May 13 upon Dennison at his house, but could never see him, and finally read a copy in the presence of Dennison's brother and servants. As regarded Tyrer, it would be proved that he was "a malt-maker, but hath no mylne to grind the same upon: which made the fraud more apparent and inexcusable in that he should grind his malt at another mill and afterwards contract for the sale of it." As regarded Dickenson, it would be shown that after the order of May 13 he did contract and contribute moneys with other of the defendants to Edwards, miller of Boughton, for him to solicit a lawyer to help them, that they might be at liberty to grind where they pleased.

On June 14, 1637, the defendants appeared for examination in the Chester Exchequer before Robert Gregg, examiner of the same court; whose signed report of the depositions of the defendants concludes with the significant remark, "The defendants are resolved to stand out against the complainant, and to free themselves if they can from being tyed and

Harl. MSS., 2083. 417.

III. DEE MILLS, CHESTER.

29. Buying Ground Malt, 1637.

bounded to grind at complainant's millnes; perceiving that the same (if they should be bounde thereunto) will be much to their damage." Some of the interrogatories put to the defendants by Gregg indicate the manner in which these Star-Chamber inquisitions extracted evidence or confession from accused persons:—

Harl. MSS., 2083. I.

Whether did you and any other of the defendants since the setting down of the said order to grind at Dee Mills meet together and resolve that you and they or any of you and whom would not obey the said order: and that you would lay your purses together to overthrow the same or lie in prison before you or they would obey the same? How many meetings have been between you, and at what place or places did you or they or any other of the defendants so meet or confer together? Declare ye truthe!

What sum or sums of money have been gathered or given by you or any of the defendants: and which of you or any other person and who by name: for the overthrowing of or concerning the said order?

How hath the said monies been employed: by whom and by whose motion was the said conclusion or attempt: and what sum or sums of money have you paid or promised or agreed to pay, and to whom, in prosecution of any course for the opposition of the said several orders of this Court or either of them?

To whom did you complain that the said order was a hard and strict order? Would you have withstood the said order if you had not been encouraged by others: and which of the defendants or what other person gave you such encouragement? Declare their names, and wrong not your conscience in concealing them, nor in covering or denying any truth which this suitor requires you to discover!

Dennison stated that he accidentally met Dickenson, one of the other defendants, and did consent to give twenty shillings to John Edwards, of Boughton, gent. (of Boughton Mill). Dickenson had informed him that he and some others of the defendants would do the like, towards bearing Edwards' charges to London to solicit help, that they might be at liberty to grind where they pleased and were best used. He had bought great quantities of malt from Edwards and his wife and from others, because he thought himself better used by them than by the millers of the Dee; "and dothe intend hereafter soe to doe if hee finde himselfe wel dealt withall therein as yet hee hath bene." He did not know any one else who did so, except with regard to oatmeal, "and excepte they bought Toll Corn at Millnes, which they receyved in meale without payinge any toll for or out of the same." It would be much more loss and damage to him to grind his malt at Dee Mills than at a horse-mill. Weaver admitted that for three years past he had not ground at Dee Mills. Formerly, indeed, he, by courtesy and for his convenience, and upon threats to take the water-supply from him, did grind there; but he received great wrong and prejudice there, and had his malt cast away with shovels before his own face, and was abused by the complainant's millers and servants.

29. Buying Ground Malt, 1637.

Harl. MSS., 2083. 450.

The result of this action also does not seem to appear among the scattered MSS. Still, as the practice of purchasing grain ready ground and importing it into a soke district was ordinarily prohibited, the Court would be compelled to decide this case, as others, in favour of the soke millers.

Text II., ch. IX., ch. XI.

30. Action against Poor Millers, 1638.

30. Concurrently with the campaign against the ground-malt ring an action was maintained by Gamull against four men—William Reeve, Thomas Reeve, William Charyles, and John Seller—who from the position of working millers had just risen to the higher dignity of lessees of a mill. The Reeves, in their appeal for a stay of proceedings, stated that on August 24, 1635, Thomas Nevitt, of London, goldsmith, demised and granted the water corn mill called Heckenhall Mill, near Chester, for twenty-one years, at the yearly rack-rent of £150, to them; and "to their understanding they became lawfully possessed thereof for the term of twenty-one years." They worked the mill for about a year, sending horses into the city for such corn as they could procure, for the most part in small and littel parceles and bagges, and

Harl. MSS., 2083. 527.

III. DEE MILLS, CHESTER.

30. Action against Poor Millers, 1638.

taking such toll as they considered lawful. Seller stated that he then took the mill from them at a heavy rack-rent for a year, but was served with process a week after; and "if he shall be debarred of liberty to grind he shall be utterly undone, being a verie pore man." They all declared that "this is a very ancient mill (claiming soke custom of its own); that there was a water corn mill where this mill now standeth time out of mind; and always the owners thereof have carried corn there out of the city on horseback and sometimes in carts, and taken toll without hindrance till now." Charyles had himself, as an employé, carried corn there for seventeen or eighteen years past. They denied they "have or ever have had in their possession any of the grants, writings, or evidences concerning Dee Mills, or other things mentioned in the bill as belonging thereto; and humbly crave this honourable Court that, pending the full hearing of the cause, they may not be deprived of the use and custom of fetching and grinding corn, for they sit upon a great rack-rent for the mill." The result of the action could only be permission to continue their trade so long as they abstained from grinding for citizens of Chester.

31. Parliamentary Ordinance, 1646.

31. Francis Gamull, who hitherto, indeed, had enjoyed little of the dignified ease popularly assigned to the lordly miller of the Dee, was destined, however, to experience still further and more serious trouble. In August 1642 symptoms of the approaching civil war broke out in Chester; and before the close of the month Charles I. visited the city, strong itself in its loyalty to the doomed king, but possessing no more devoted servant to the royal cause than Francis Gamull, who took a foremost part in the military duties which fell upon the civic rulers of the city. He had the honour of entertaining Charles

31. Parliamentary Ordinance, 1646.

in his house in Bridge Street, near the mills, a mansion which, in part, still exists; was knighted; was with Charles at the Phœnix Tower on the walls, and saw thence the rout of the royal forces at Rowton Heath; accompanied, with five hundred horsemen, the retreating king on his departure from the city, September 1645, across Dee Bridge, under the walls of the ancient mill, into Flintshire; and at the

Antique View of Dee Bridge and Mills.

imminent fall of the city was first of the commissioners of defence to sign a protest against the articles of surrender, February 1646. Truly, when the end came, he had done enough to render himself a marked man among the Parliamentarians, and in due course ensued the inevitable result. Meanwhile, his discomfiture was the opportunity of the citizens; and these, with the mayor at their head, appealed to the Commonwealth for the abolition of the causeway and the mill. In various letters on this subject (still existing in the city archives) passing between the mayor and the parliamentary representatives of Chester is fully evidenced the determination of the

III. DEE MILLS, CHESTER.

31. Parliamentary Ordinance, 1646.

city to take advantage of circumstances so favourable to their views, and ultimately an order was issued for the destruction both of mills and causeway, and the erection of new mills by the town on the Roodeye:—

Die Iovis Primo Octobris 1646.

Whereas the city of Chester hath from the beginning of these unnatural wars been under the power of the enemy: and many aldermen, sheriffs, peers, and common councilmen of the said city have taken up arms, &c., against the Parliament, &c. . . .

Be it ordered and ordained by the said Lords and Commons . . . that the milnes called Dee Milnes, near the said city, and the stone causeway erected for the use of the said mills, holden by Francis Gamull, alderman of the said city, a Delinquent, being a great annoyance and obstruction to the trade of the said city and otherwise very inconvenient, be demolished and taken down within the space of four months now next ensuing at the charge of the said city.

Harl. MSS., 2057. 10, 54.

And that in the meantime the profits and emoluments of and belonging to the said milnes shall be to the mayor of the said city for the time being for the support of his office.

And that within the space of one year there shall be made new watercourses for other milnes to be erected upon a parcel of land called the Rood Dee, whereunto the materials of the said old milnes and causey shall be employed: the city to pay fee-farme rent as was payable hitherto.

Provided always the said Francis Gamull shall be admitted by both Houses of Parliament to make his composition for his delinquency: in which case he shall receive satisfaction for the said mills out of his fine.

The severe order was never carried out, though on December 6, 1648, the corporation prepared for the destruction of the causeway, as appears from an entry in the Assembly Book of the city:—

Ordered that Thomas Prenton shall be the principal workman to work, and to order work to be done by labourers and others that he shall take to his assistance, with the advice and consent of the overseers, to take down and demolish the causey near to Dee Mills, which by the last Assembly was ordered to be immediately taken down in pursuance of an Ordinance of Parliament which declared the said causey to be a nuisance to the city.

Again was the order evaded, and the causeway, like the mills, was left intact. Sir Francis Gamull, who must have exerted very considerable influence to secure this result, died November 27, 1654, and was buried at St. Mary's. One of his sons had died young, and the other had been slain in the war; and his property passed to his daughters. "The several shares in the Chester mills subsequently experienced many successive alienations"; but the soke by this time was irremediably broken, and the mills no longer enjoyed feudal rights nor exacted feudal servitudes.

31. Parliamentary Ordinance, 1646.

32. The later history of the establishment may, in conclusion, be briefly noted. Early in the eighteenth century the place became the property of Edward Wrench, who obtained permission from the corporation to add a new bolting mill at a rent of 12d. per year (October 1747). This consent was followed three months later by the order:—"That the lower window of Mr. Wrench's new bolting mill on the bridge be stopped up, it being apprehended that the reflection of candle light through the same at night will be dangerous to travellers on horseback passing the same"; a decision which contrasts with the custom at Wakefield, where in the window of the chapel of St. Mary on the bridge was constantly exhibited a light for the guidance of travellers. On September 26, 1789, the mills were burnt down. On April 23, 1790, Mr. Wrench obtained the consent of the corporation to rebuild and "extend the mills upon the western part of Dee Bridge on payment of five shillings, including the one shilling per annum now paid for a previous like liberty." On August 3, 1807, they were advertised for sale in the *Liverpool Advertiser*, being then stated to contain "twelve pair of stones, six of which are French." On March 6, 1819, they were again burnt down; and on January 6, 1847, were a third time similarly destroyed. They subsequently passed into the hands of Alderman William Johnson, who introduced rollers in place of the ancient stone system. On May 29, 1895, they were for the fourth time wrecked by fire, the central portion of the structure being burnt out; and in April 1895 the mills, together with what water rights they possessed in virtue of the historic causeway, were purchased by the corporation.

32. In Modern Times.

TOWNSEND, THE LAST OF THE KING'S MILLS OF LIVERPOOL.
View from Mill Lane, Islington.

See page 182.

CHAPTER IV.

KING'S MILLS OF THE ROYAL BOROUGH OF LIVERPOOL.*

IV. KING'S MILLS, LIVERPOOL.

1. Eastham Watermills, 1257–1423.

1. THE archæology of corn milling in Liverpool widely contrasts with that of any of the medieval cities already considered. The royal soke of the manor, small and insignificant enough when first discerned in Plantagenet days, grew concurrently with the development of the embryo town and port; and safely endured beyond that critical period which extinguished the soke of Dublin Castle and Shrewsbury Abbey, though it was never comprised in one powerful mill, as was the soke of Chester in Dee Mill. The king's mills of Liverpool consisted of several separate establishments, held directly under the Crown, usually conjointly; and sub-leased separately. So long as the soke was retained by the Crown, these mills successfully conserved their legal rights; but when that binding link was broken—this being when Charles I. in 1629 alienated his ancient inheritance in Liverpool and its mills—the new holders of the soke

* By the favour of the Corporation of Liverpool, and many obliging facilities of officials in the Town Clerk's department, much information relating to the milling history of the town has been derived from the Port Mote books, dating from the year 1551, and the Okill collection of transcripts of early Duchy deeds, &c., included in the archives of the city.

A brief outline sketch, *The King's Mills of Ancient Liverpool*, prepared from the draft of the present chapter, and contributed by Mr. Bennett to the Historic Society of Lancashire and Cheshire in 1896, varies in several material details from the history now presented. This is due to many further investigations having since been made to solve difficulties, then experienced, in linking together the fragmentary records which are all that remain to Liverpool. A fair amount of success has resulted; mainly by the kindness of Mr. R. D. Radcliffe, M.A., F.S.A., and Mr. W. Farrer, in affording full investigation of transcripts of the Moore and Crosse Charters, and of many local deeds found by the latter gentleman in Additional MSS. at the British Museum.

IV. KING'S MILLS, LIVERPOOL.

1. Eastham Watermills, 1257–1423.

were unable to resist popular encroachment. Little by little it was thus frittered away, chiefly by the aid of the corporation; till finally, greatly to the loss of the owners, whose rights were not bought but overridden, it was destroyed.

Nothing is known of Liverpool mills, and very little of the town itself, before the year 1257, the starting-point of local milling history. In many places throughout the kingdom we may point with certainty to mills being owned and worked, at the earliest period of our national industrial history, by Saxon thanes and freemen; but no such retrospection may be made at Liverpool. It is, however, impossible to avoid the conjecture that the town—a chartered port, and seat of a royal castle, in 1206—possessed some kind of corn mill other than querns earlier than 1257; and that at that date the avocation of the dusty miller was already established as one of the first manufacturing industries of the place. A horse-mill would doubtless have been set up in the castle; probably at the same "bakehouse," which subsequently was included in the *Extent* of Liverpool in 1347, was repaired in 1476,* and kept in use till about 1670. One or more watermills also would, no doubt, have begun to cope with the necessities of the increasing population. One point, however, is certain; the charter by King John (Earl of Morton) in 1206 had made no allusion to milling soke, and therefore, as lord of the manor, he had tacitly reserved it.†

Text, II. 207.

Duchy Orders, Edw. IV. 100.

* 1476. Writ for repair of Liverpool Castle: "The walles of ye Bakehous within the Castell to bee taken down for reparacon of ye sd tower."

† John's charter to Liverpool is of the same type as the one granted to the town of Bristol. That neither of these allowed exemption from milling soke is not alone evident from subsequent events, but may be directly proved by reference to another charter of the same Bristol type, in which exemption from soke is distinctly granted by a special clause; viz. the charter to Lancaster in 1193, which stipulates, "Furthermore, I have declared the said burgesses freed from suit to my mill."

Arch. Journal, 1898, 359.

IV. KING'S MILLS, LIVERPOOL.

1. Eastham Watermills, 1257–1423.

Ere long the manorial rights had passed to Robert de Ferrers, Earl of Derby, a minor in ward to Prince Edmund, son of Henry III.; and in the *Extent* of the manor for 1257 Henry de Lee, bailiff, accounts for £10 for the fee-farme of the town of Liverpool, including "two watermills and one windmill." On the forfeiture of the Ferrars' estates Henry conferred them, including Liverpool, on Edmund Crouchback, by whom the three mills were leased out to some holders, not now known, but very probably the general body of the burgesses; the milling soke of Liverpool being thus established, as well as the system of leasing and sub-leasing which continued to prevail so long as the soke existed.

Beyond this definite point the records of the watermills—to which alone it is first proposed to refer—are extremely brief and vague:—

1297. Inquisition on the death of Edmund Plantagenet. The town contains "two mills, one a watermill and the other a windmill, worth by the year 5 marks [£3 6s. 8d.]"; the entire rental of the town being £25 10s. — Chancery Rec., Inq. p. m.

1326. Inquisition on the death of Thomas, Earl of Lancaster. The manor possesses "one windmill of 26s. and one watermill of 24s. per annum."

At the inquisition on the death of Earl Henry, in 1361, the estate is found to comprise "two windmills and one horse-mill"; and mention of the earl's watermills does not again occur.* It seems, however, that they had not been destroyed but merely sold, probably as old-fashioned and inefficient; various

* In this connection may be mentioned a defaced, and possibly misleading, deed, comprising a rental of Liverpool lands; which, though undated, proves by an examination of the names of tenants to be of about 1330–32. It mentions certain land held by "— del More in le Bonkstrete [Water Street] nup' ad. molend."; apparently referring to land formerly attached to a mill in that street. But there appears to have never been a mill in le Bonk Street; and the allusion in the deed is to one Adam Molendinarius, who is found granting land in le Bonk Street and elsewhere in 1332 to John, son of William del More, and others. He may no doubt be identified with the De Molines family (whose names appear in various early undated deeds), ancestors of the Molyneux family, from whom descend the Earls of Sefton. — Moore Deeds, 100, 100[a], 106[a]. [Morton's numbers to these deeds are those quoted in the whole of these references.]

IV. KING'S MILLS, LIVERPOOL.

1. Eastham Watermills, 1257–1423.

incidental allusions appearing to prove their continuance as private mills for some time. The following refer to one of them, viz. the Castlefields Mill :—

Moore Deeds, 262 (3). — N.D. (*c.* 1257). One selion in Le Lendingfeld * [field containing the landing-place on the pool], with a quarter burgage in Le Mulnefeld.

Ibid., 101. — 1334. Certain butts of land in Le Lending extending from Le Mulne rakes [mill fields] to the bank [of the pool].

Add. MSS., 32105. 411. — 1387. Land near Le Milnedame.

Crosse Deeds, 92. — 1401. A burgage in Le Dalestreete bordering on the king's highway [Kirkdale and Scotland Roads] † and extending to Le Milnedame.

Add. MSS., 32105. 579. — 1420. A messuage in vil de Lyverpull, viz. in le dalestreete between the tenement of Robert de Derby and le Milne dame.

Ibid., 32105. 574. — 1423. Sciant quod ego Henricus de Bretherton capellanus dedi Nicholas de Lyverpull clīco et Johannė uxori eius et hered inter eos exeuntibus omnia illas terras [&c.] que habui ex dono et feoffamento eiusdam Nichās in campis et villa de Lyverpull. . . . Item quinquaginta pedes in latitudine et quadraginta pedes in longitudine in le Castelfeld ex dimissione Ric̆hardi Chenall ubi molendinū aquaticum stat. . . . Item except uno burgage in le dalestreete et except uno molendino aquatico juxta Dalefield. Dat apud Lyverpull die Sabbi in ffesto Sancti Mariæ Magdalene, anno 5 H. 5.

Be it known that I, Henry de Bretherton, chantry priest, have granted to Nicholas of Liverpool, cleric, and Joan, his wife, and their heirs, all those lands, &c., in the fields and town of Liverpool which I have had by gift and feoffment of the said Nicholas. . . . Item, a plot 50 feet by 40 feet in the Castle Field, demised by Richard Chenall, where the watermill stands. [With reversions] . . . except one burgage in Dale Street and one watermill juxta the Dalefield. Given at Liverpool, Saturday of the Feast of Mary Magdalene [July 22], 1423.

Crosse Deeds, No. 1. — * N.D. (*c.* 1260). Land in le Lendingfeld extending from the sea to the moor :—a mare usq le mor.

Moore Deeds, 509. — N.D. (*c.* 1290). Land extending de via regia usq' Le Poledych [the Pool].

Ibid., 187. — 1361. Land in Le Lendyngh.

Ibid., 333. — 1507. One butt end in le Poledych in the field of Liverpool, on which is the ancient fosse or ditch with the merestones or boundary-stones :—super quid'm fosset cum sepe merestans est de antiquo.

Ibid., 350. — 1524. Two sandlands juxta the White Crosse, abutting at the east on Le Nether fylde de Eu'ton, and extending beyond the way which runs between Walton and Liverpool, on Le Poledych.

Port Mote, i. 40. — 1541. No person to lay any ballast or rubbish in the Lake upon the Warthe, alias the Wharfe.

Crosse Deeds, 16. — † 1305. Lands in le Dalefield juxta viam regali.

Add. MSS., 32105. 574. — 1423. Lands in Erstandale bordered by regia via de Kyrkdale.

IV. KING'S MILLS, LIVERPOOL.

1. Eastham Watermills, 1257–1423.

In these references appear to be indicated the site of the mill which had disappeared from the manorial schedule by 1297. It is stated to be in the Castle-fields, which extended from the Castle Hill down to the borders of the creek or pool.* It is also said to be in the Dale Street and Field; and was therefore situated at the Castlefields' side of the pool, near the foot of Dale Street (about opposite to the present Technical Schools in Byrom Street). Here presumably were the landing-place and the mill fields

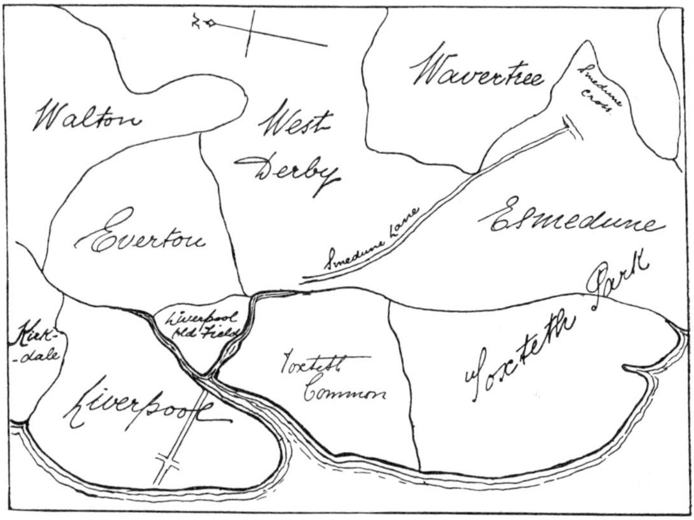

Lord Derby's Map, showing Liverpool Old Field.

of 1257–1334, the mill-dam of 1387–1420, and the mill itself, mentioned in the deed of 1423. The Castlefields Mill has not been mentioned before in local archæology, and it has been supposed that the two original mills stood both together at the place now to be mentioned.

The second of the mills was situated alone in the fields below the heath on the opposite side of

* "Le Castlefeld" is mentioned in 1330; and "Land in le Castelfeld abutting upon le Pulle" in 1372. The name seems to have become obsolete in the fifteenth century. Moore Deeds, 106a, 222.

IV. KING'S MILLS, LIVERPOOL.

1. Eastham Watermills, 1257–1423.

the pool; not actually on the pool itself, but upon a small brook which there debouched into it. The land on that side of the pool comprised the dale of the small hamlet and manor of Everton.* But before 1314 one of the earls, solicitous even then for a "greater Liverpool," had transferred a large portion of this land from the liberties of Everton to those of Liverpool; hence Everton Dale was in the territory of Liverpool.† The annexation seems to have been made by Henry, first duke, in 1309, when he granted to the burgesses a certain ten acres of land adjoining the side of the town; ‡ this being apparently the larger or "mickle" Old Field of Liverpool.§ The mill may have been there when the land was annexed to Liverpool; and, from that circumstance, or from the fact of its site being in the old Everton Dale, it was known as "Everton Mill"; as, in matter of fact, its successor on the same site (a windmill) also was. In any case, the old watermill of Everton, in the enlarged territories of Liverpool, really was a Liverpool mill. From that date, so far as the records show,

Text, I. 155.

Sheriff's Accounts, 7 Hy. VII. to 4 P. & M.

* At Everton was conserved, till towards 1516, as a source of manorial income, the licensing of millstone makers. In the early days of difficult and costly carriage, local stones were of course used for hand or power mills wherever obtainable; and for digging or quarrying them lords of manor ordinarily exacted a rent. At Everton in the time of John of Gaunt mention occurs of a licence for the millstone quarry; and in 1516 Sir R. Molyneux, Duchy Receiver for that manor, includes in his accounts the same old licence quærendi petres molares in quærra ibidem; though "nihil" was received for the same. The stones available at Everton could but have been boulder-stones from the drift overlying the red sandstone formation of the district, and useful probably only for querns.

Add. MSS., 32105. 12, 129. Moore Deeds, 113.

† 1314. Four butts lying on Euerstandale Banck in the territory of Liverpool.
1322. Lands on Le Knolle [Everton Brow] stretching towards Euerstandale.
1338. Lands in the territory of Liverpool in Euerstondale.

Okill, West Derby MSS., iv. 161.

Ibid., iii. 205.

‡ In the possession of the Corporation is a copy of a grant signed by Henry at Liverpool in April, 3 Edward II., of "six" acres of heath-land "encostaunts la quoté de la ville"; the word "six" seeming to be a clerical error for "dix." The grant of "ten acres" is referred to in various other deeds; and Okill, in reproducing a map of the district, has marked in the margin that the ten acres were in this part of the Old Field. Respecting the map, a copy of which appears on p. 129, he remarks, "The situation of the Old Field is shown in a plan obtained from the Earl of Derby, and now preserved at Knowsley."

Moore Deeds, 142. 124.

§ 1346. Le Lytil Oldefeld in Lyverpull.
1349. Le Mekiloldefeld in Lyverpull.

IV. KING'S MILLS, LIVERPOOL.

1. Eastham Watermills, 1257–1423.

Everton was left entirely without a mill of its own; a circumstance doubtless due to the fact of its district being completely surrounded with rural mills, all within reasonable distance.* In closest proximity was the Liverpool mill called "Everton Mill," and farther afield were Kirkdale Windmill,† Bootle Watermill,‡ Spellow Windmill,§ Skelmersdale Watermill,|| and the

* In a series of old charts of river-approaches in the possession of the Mersey Docks and Harbour Board (inspected by the courtesy of the Board) are two by Fearon, dated 1736, upon which a tower is marked as "Everton Mill." This, however, was the tower of Everton Beacon, which was blown down in 1803.

Everton Beacon in 1798.

† Existing in 1317. Leased out in 1340 by Alice, widow of Robert, Lord of Kirkdale. Text, post, p. 161.

‡ Quoted in deeds temp. Edward III. Ibid., post, p. 160.

§ See Appendix to this Chapter.

|| Leased out by Sir Thomas Gerard, 1412:—

Hic indentā testat^r q^d Thomas Gerard miles concessit & ad firmam demisit Ricō de Assheton de Skelm'sdale & Agnes vxī eius Robto & Iohi filijs eoz Molendinum aquaticū de Skelm'sdale cū stagno & refullo aque eiusdem Molendinum & cū una p'cella t're nup' in tenura Rici de Asshehurst iacente int' p'dtm Molend & le ffendour [?]. H'nd & tenend p'dtm Molend cū p'dict stagno refullo aque p'cella t're & cū racōnabili introitu & excitu p'dictis Rico Agneti vxi eius Robto & Iohi a festo Natal dnī Anno regni Regis Henr quarti t'tio decimo usqz ad finem triginta annoz extunc px sequens: Reddendo inde annuatim p'dto Thome hedibz & assignatis suis nonem solidos & duos denarios argentī ad festu Nativitate S'ti Ioh'is Baptē & S'ti Martini in yeme p' equales porcōnes. Et p'dti Ricus Agnes Robtus & Ioh'es p'dta Molendinū & stagnū cu p'tin b'n & competent rep'abunt & sustentabunt sumptibz suis p'pis durante t'mino p'dto et in fine t'minū p'dti in competenti statu ea dimittent: et p'dti Ricūs Agn Rob & Ioh'es concedunt p' p'sentes q^d si contingat eos om'es infra t'minu p'dtm obire tunc liceat p'dto Thome hedibz & assign' suis in p'dta Molend' stagnū refull' & p'cell' t're cu p'tm reintrare & in pristinis statu suo ea retenere impp'm. Et ecia si contingat p'dtm redditū aretro esse in p'te vel in toto ad aliquē t'minū p'dtm tunc b'n liceat p'dto Thome hedibz & assign' suis in p'dtis Molend' refullo stagno & p'cella tre cū p'tin distring'e & districcōnes ibidem captat inde fugare imp'cere & retenere quousq' de p'dtu redditū cū arreragijs eiusdem si que fu'int eidem Thome hedibz aut assignatis suis plenarie fu'int satisfactu. Et p'dtus Thomas & hed'es sui in p'dta Molend' stagnū refullū & p'cell t're cū p'tm p'fatio Rico Agneti Robto & Ioh'i & unicuiq eoz diutius viventi usqz ad finē t'minū p'dti si ita dni vix'int contr' om'es gentes warantizabunt. In cuī 'rei testimoniū p'tībz hui' indentur p'dto sup' d'te sigilla alt'natim apposuerunt. Moore Deeds, 735ᵃ.

Dat apud Assheton in Makirfelt die Lune in festo Assumpcionis b'te Marie Virginis anno regni Regis hen^r quarti post conquestū Angl' t'tio decimo.

This indenture witnesseth that Sir Thomas Gerard has conceded and demised to farme to Richard de Assheton of Skelmersdale, Agnes his wife, and Robert and John his sons, the watermill of Skelmersdale, with the pool and water-race of the said mill, and a parcel of land formerly in the tenure of Richard de Asshehurst

IV. KING'S MILLS, LIVERPOOL.

1. Eastham Watermills, 1257–1423.

Liverpool rural windmills at Accers, Wavertree, and West Derby.* The site of this old Everton watermill was the only one possible over a radius of several miles; so that little hesitancy need be experienced in endorsing the identity claimed for it in the unpublished MSS. of the local antiquary Okill. The Moss Lake stream coursed across the heath and down a steep declivity in the daleside; and at this spot the pool and dam were constructed and the mill erected; the stream a few yards farther (near the junction of the present Richmond Row with Scotland Road) running into the pool nearly opposite the Castlefields Mill. It is this mill which seems to be alluded to in the inquisition of 1322; and its dam or dyke in two later deeds:—

Okill, West Derby MSS., iii. 205.

Maps, pp. 129, 135.

Moore Deeds, 111.

1337. One selion in a certain field in Liverpool called the Dykfeld de Eu'ton.

Add. MSS., 32105. 206.

1358. Four selions in the town of Liverpool near Le Dckfeild de Euerton.

No direct reference to the mill seems anywhere to occur; and though in 1384 we read of certain land "juxta Eu'stan Mylne," the mill in question was no doubt the windmill which before that date had

Moore Deeds, 259.

lying between the said mill and le ffendour:—To have and to hold the said mill, with pool, water-race, and land, and with reasonable entrance and exit, from Christmas, 13 Henry IV., for thirty years thence following; paying yearly to the said Thomas, his heirs and assigns, 9s 2d. in silver at St. John the Baptist's and St. Martin's in Winter in equal portions. And the said Richard, Agnes, Robert, and John shall well and competently repair and sustain the said mill, with pool and appurtenances, at their own proper costs during the above term, and at the end shall deliver up the same in good condition: and the said Richard, Agnes, Robert, and John have conceded by these presents, that if it happen all of them should die during the said term, then it shall be lawful for the said Thomas, his heirs and assigns, to enter and possess and retain the same in perpetuity, as before this grant. And if the said rent be wholly or in part in arrear at any term, then the said Thomas, his heirs and assigns, shall distrain upon the said mill, &c., and take sufficient from such distraint to fully satisfy all such arrears. And the said Thomas and his heirs shall warrant the said mill, pool, water-race, and parcel of land to the said Richard, Agnes, Robert, and John, or whichever of them by the will of God live longest, against all men, till the end of the said term. In testimony of which the parties to this indenture have alternately fixed their seals.

Given at Ashton-in-Makerfield, Monday of the Feast of Assumption, Henry IV. after the Conquest, thirteenth year.

Text, post, p. 184.

* These appear in Liverpool leases from 1475.

IV. KING'S MILLS, LIVERPOOL.

1. Eastham Watermills, 1257–1423.

been established there, and which quickly minimised the importance of the watermill and in due course entirely superseded it. But though the latter disappeared, its pool and dam, or dyke, in the Everton Dale long remained, and are frequently mentioned in the town records. In the meantime the designation of the Liverpool portion of Everton Dale gradually became transformed from "Everton Dale" to "Eastham Dale" or "Eastham"—an ancient and once celebrated place-name now revived by its milling records.*

2. Eastham Watermills. Reliquiæ.

2. Eastham, standing out on the open heathlands beyond the town, with its brook coursing down beside the busy mills that centred at this spot the milling trade of Liverpool for several centuries, was then, no doubt, a pleasant-enough place; where oft were witnessed—

> ". . . The brimming wave that swam
> Through quiet meadows round the mill;
> The sleepy pool above the dam,
> The pool beneath it never still;
> The meal sacks on the whiten'd floor;
> The dark round of the dripping wheel;
> The very air about the door
> Made misty with the floating meal."

But after the decay of the watermills, the little stream flowed useless and neglected below the huge king's windmill that long stood there; and in 1561 it is to

* N.D. (*c.* 1257).	Land in the territory of Liverpool between the Crosses juxta Eu'ston.	Moore Deeds, 269 (8).
1385.	Land lying juxta Kirkdale between the White Cross and Eu'ston Crosse.	Ibid., 258.
1410.	A sandland juxta Euerstan Milne.	Add. MSS., 32105. 469.
1451.	Lease of Etton Mylne.	Text, post, p. 142.
1521.	A sandland which shoots from Estam Mylne towards the Dale.	Moore Deeds, 90.
1563.	The lane that leadeth from St. Patrick's Crosse into the townfield towards Eastham Mylne [Great Crosshall Street].	Port Mote, i. 149.
1563.	The town's land called Mylne Dam, East-town.	Ibid., i. 192.
1564.	Le Eastham Dale in Lyverpole Field.	Crosse Deeds, 195.
1566.	Three butts in Estam Dayle in Lyverpole.	Ibid., 205.
1567.	One sande lande and two butts shooting in to Estham Dale belonging, to the said Our Ladie Hall.	Port Mote, i. 386.
1577.	A stone coppe or ditch, being near unto Eastham Milne.	Ibid., ii. 144.
1587.	Comonly called Eastham Milne, within her majesty's manor.	Text, post, p. 162.

IV. KING'S MILLS, LIVERPOOL.

2. Eastham Watermills. Reliquiæ.

Lanc. and Ches. H. S., 1851, 16. Port Mote, ii. 542, iii. 422.

be recognised under the sorry cognomen of "the long diche called the common diche upon the sowth p'tie of the galow felde"; a designation which occurs again in 1599, when John and Richard Crosse were ordered to secure their "ditch in Eastham mylle lane"; and in 1647, when "the diche in Eastham mill lane" once more came under the notice of the corporation.* In 1587 it was pleaded, in an action brought on behalf of Queen Elizabeth's soke mills, that "her Majesty hath no watermilns in Liverpool," which indeed was the case, though there were one or two illicit ones at work in the town. The Eastham Mill Dam, with its croft, still remained, however, and was let by the corporation for grazing purposes; in 1541 Adam Dandye paying "for milnedam x[d], for mylne dale xviij[d] per annum"; and in 1563 the place appearing as "part of the townes land called Mylne

Ibid., i. 61.

Ibid., i. 192.

Crosse Deeds, 151.

* As early as 1476 a portion of the field by the mill-stream appears as Le Galowffeld in camp de Lyverpull; and ages after being relieved of its ancient use this field frequently appears under its old gruesome title; even as late as 1753 we may read of "the field at the bottom of Dale Street called the Gallow Field."

Port Mote, x. 587.

The Liverpool authorities seem to have been much impressed with the suitability of the vicinage of corn mills as sufficiently public places for the erection of their "galows"; and just as was one of the earliest, so was one of the latest of the local mills endowed by the authorities with this doleful association. On July 2, 1715, the corporation granted to Mr. Gray a lease of land "in the lane leading by Mr. Norris' mill towards the gibbet to build a mill." Norris' mill was a watermill built outside the liberties of the town some little time before 1587. The new mill of Gray became known throughout the eighteenth century as the Gallows Mill, and here shortly before its erection four of the victims of the Jacobite rising of '15 were executed. "The captured prisoners were crowded into the gaol at Liverpool, and in January 1716 no fewer than 113 of them were set down for trial in this town by a Commission comprising Baron Montague, Baron Bury, and Mr. Justice Eyre." The corporation, probably remembering their Vicar-of-Bray-like tergiversations of about half a century before, promptly held a special council meeting January 6, and loyally offered the honour of the freedom of the city to Sir T. Bury, Sir R. Eyre, and Sir J. Montague, with other legal lights who had accompanied them to the trial; but had the mortification of being compelled to enter in the margin of the record in their book that "the Judges refused to be sworn." In the meantime the trials progressed. "Before the end of January seventy-four of the prisoners were condemned to death, and ordered to be executed in different parts of the kingdom. None of them were natives of Liverpool, and Liverpool had been in no way implicated in the rebellion; but the judges thought fit to order four of the condemned to be executed in the town—Alexander Drummond, gent., Scotsman; Archibald Burnett, of Carlops, gent., Scotsman; George Collingwood and John Hunter, both of Northumberland. On February 25 these executions were carried out with all the barbarous practices of quartering and burning, and were the last executions of the kind in the kingdom." Two additional mills were subsequently built here.

Ibid., viii. 125[b].

App. to this Chapter.

Port Mote, viii. 131[b].

IV. KING'S MILLS LIVERPOOL.

2. Eastham Watermills. Reliquiæ.

Dam, East-town [Eastham]." The dam or "causeway" appears in a rental of about 1545—"one land at Euerton cossy, two lands shooting upon eastm mylne, two lands in the Galow felde"; and again in a conveyance of 1549—"a croft nygh Eu'ton causey.' In 1708 some reminiscence of the Eastham Dale of 1314 may be traced in the Easting Dale of the poor-rate book;* and in 1720 a sketch-map belonging to the Earls of Sefton, former lords of the manor, shows the diminished brook still running to the "East-tyn Myll Dam." But towards the close of the eighteenth century the drainage of the uplands at length destroyed the stream; and in 1790 and 1807 the maps

Moore Deeds, 108, 363.

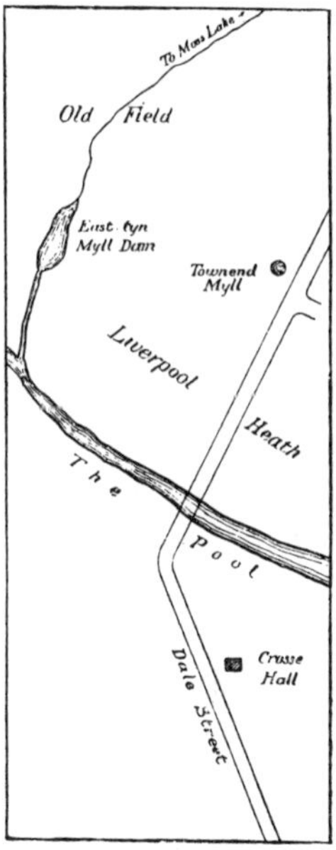

Eastham Mill-pool and Townsend Mill, 1720.

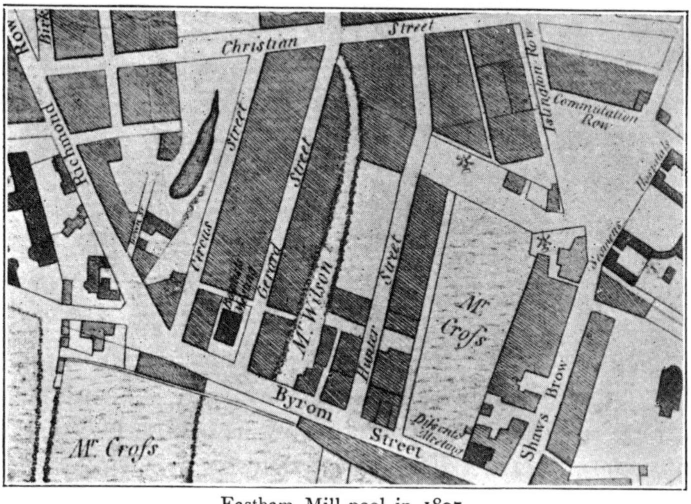

Eastham Mill-pool in 1807.

* "List of lands and enclosures in Liverpool: extracted from the poor-rate book for the parish, now [c. 1750] in the possession of Henry Holmes, Esq." Okill MSS., iv. 526.

IV. KING'S MILLS, LIVERPOOL.

2. Eastham Watermills. Reliquiæ.

of Liverpool show the pool alone—overlooked by the south side of Circus Street, the van of an approaching phalanx of streets that shortly overwhelmed both it and the entire countryside.

At the present day the site can still easily be identified near the rear of the Art Gallery, where may yet be seen, in an open space recently cleared of a number of old houses by the corporation, in Down Street (the ancient Eastham Mill Lane), the sloping declivity which once held the pool backed up by the dam. Though in the midst of surroundings little suggestive of ancient rural beauty or industrial prosperity, and though not otherwise fascinating, the site of medieval Eastham—the very name and existence of which have for so long been forgotten—may still claim an interest of its own as the Plantagenet birthplace and medieval centre of corn milling in Liverpool.

3. Eastham Windmill. Leases, 1257–1413.

3. A windmill succeeded the watermills at Eastham, and till the eighteenth century the neighbourhood remained the site for most of the windmills of the town. The allusion to the watermills in 1257 comprises also mention of an existing windmill; this being only sixty-six years after the date of the first known windmill of Europe. It is seen to be still existing in 1297; and again in 1326, when it is valued at 26s. as against 24s. per annum for the remaining watermill. It was doubtless "the mill of Liverpool" for the repair of which, in 1342, the verderers of the Duchy Forest of West Derby reported that two oaks had been cut down in the time of Gilbert of Haydock, seneschal. Though no indication yet appears as to site, its identity may definitely be decided as that of the windmill at Eastham.

Text, II. 235.

Inq. Forest, 16-17 Edw. III. 8.

Close Rolls, Duke Hy., No. 3.

In 1357 several representative burgesses received from the Duke a lease of "our town of Liverpool

SITE OF MEDIEVAL EASTHAM, LIVERPOOL.

3. Eastham Windmill. Leases, 1257–1413.

and all our mills of the said town." These mills are to be identified as those of Eastham and Townsend respectively. Some years after 1357 one of the two, that of Eastham, seems to have been acquired, either on a leasehold or a copyhold tenure, by one or more of the burgesses; being, in 1375, in the possession of Richard Munn, priest, and John Henthorn; who then re-conveyed it to William the son of Adam of Liverpool.* Its owner, one of the co-lessees of town and mills in 1357, was a prominent public man of his day, several times mayor, and, now, the first known mayor of Liverpool on record, (4 April, 26 Edward III., 1352).† Richard Munn's deed of conveyance will be noted to state the site of the mill in 1375:—

Moore Deeds, 450.

Sciant psent & futuri q[d] nos Ioħ henthorn & Ricus Munne capeƚƚ dedim[s] concessim[s] & hac pn̄ti carta nr̄a confirmavim[s] Witt fiƚ Adē de Lyv̄pull & Kat'ine vxi eius. . . . Unū molendinū ventriticū & tres seliones in le Micilholdefeld sup̲ quas p'dtim molendinū stat iacent int' t'ra Ioħ Carlele & t'ra Rad de Denbegh. . . . Ecia unā placeā t're in le Boncstrete iac int' ten' scī Nic̄ et teñ Ioh. de Stanay continent viginti pedes edificat & nonē pedes nō edificat in longitudiē et octodeci pedes in latitudiē. Quē quidē mesuag' t'ras & teñ cū molendino ventritico piscar turbar' pastur' cū omībus p̲'tm suis huimus de dono & feoffamento d'ti Willi fil Adē.—Dat Ap[d] Lyv̄pull quinto die Martij a[o] r. r. Edwardi t'tij a conquestū quadragesimo nono.

Be it known to present and future that we, John Henthorn and Richard Munn, priest, have given and conceded, and by this our Ƿresent charter have confirmed, to William the son of Adam of Liverpool and Katharine his wife . . . One windmill and three

---

* Duke Henry, in 1361, had granted William Fil Adam a rental of 20s. per annum from the manor; and this may have been the origin of his milling estate.

Lanc. and Chesh. Hist. Soc., 1896.

† Our earlier investigation of Eastham milling disclosed William de Grenolph (June, 27 Edward III., 1353) as a predecessor of the mayor of the year 1356, thitherto accredited as the first mayor; the present researches, however, giving William Fil Adam a still earlier precedency. See *The Chapel of St. Mary del Key, Liverpool;* and *Early Recorded Mayors of Liverpool* (Elton).

Ibid., 1902-3.
Moore Deeds, 257. 112, 119[a].

In addition to the mill, William also held a soke-oven in the town; a quit-claim of 1383 mentioning his "p'prius pistrinum" in Castle Street. This bakery had apparently been there in 1337, when a deed by Adam of Liverpool (father of the above Wılliam) mentioned the tenement of the heir of Robert Pistor near the house of the constable of the castle.

IV. KING'S MILLS, LIVERPOOL.

selions of land in the Mickle Oldfield upon which the said windmill stands, lying between the lands of John Carlisle and Richard Denbigh. . . . Also a plot of land in the Bank Street,* lying between the tenement of St. Nicholas' Chapel and of John de Stanay, comprising 20 feet built upon and 9 feet not built upon, by 18 feet; which messuages, lands, and tenements, with windmill, fishery, turbary, pasture, and all appurtenances we have had by the gift and feoffment of the said William the son of Adam. Given at Liverpool, fifth day of March, in the forty-ninth year of the reign of King Edward, the third after the Conquest [1375].

3. Eastham Windmill. Leases, 1257–1413.

This ancient windmill, with its three selions of land, stood in the Mickle Oldfield, which, with the Little Old Field, is well accredited as having been on the town lands at Eastham. The practice ordinarily adopted when windmills were first erected was to build them at the usual milling locality near the watermills; and it is quite in accordance with this practice that we find the first windmill of Liverpool erected at Eastham, on the open breezy heath above the dam and pool of the declining watermill.† William Fil Adam died in 1383, and in 1384 a passing allusion to the mill in one of the Moore Deeds—land in le Mickleholdefeld juxta Eu'stan Mylne—seems to be its first recorded designation.

Text, II., ch. xv., § 8.

Crosse Deeds, 77.

Moore Deeds, 259.

In 1390, another deed also testifies to the situation of Eastham Mill in the Mickle Old Field:—"one selion of land juxta Eu'stan mylne, one selion in Le Mekelholdefeld, and one selion in the said field extending to the mill: in campo p'dto vsq' molend." In 1394 John of Gaunt renewed the lease of the town to the corporation, "together with all our mills to the said town belonging," with the usual stipulation

Crosse Deeds, 83.

* N.D. (*c.* 1257). Land in a certain street called Le Bonkestrete.
1560. Bonke Strete, nowe called the Water Strete.

Moore Deeds, 271 (11). Ibid., 6.

† This high ridge of land, after the suppression of milling soke at the end of the seventeenth century, was utilised for the erection of no fewer than nine windmills, extending southward from Eastham. Nearest to the latter were two at the top of Shaw's Brow and Clayton Street, on the site of the Art Gallery, and closely adjoining the old site of Townsend Mill: the others were the mills of Lime Street and Copperas Hill.

App. to this Chapter.

IV. KING'S MILLS, LIVERPOOL.

for repairs; the same being renewed by Henry IV. in 1401:—

3. Eastham Windmill. Leases, 1257–1413. Duchy Leases, 1-4 Hy. IV.

Ita semper quod iidem Thomas Robertus Ricardus & Wills facient & reparent molendina pdca sumptibus suis propriis capiend' sufficiens maeremio ad molendina pdca facienda & reparanda quotiens opus fuerit in parcis nostris de Toxstath & Croxstath & bosco nostro de Symondewood p visum & liberation' forasteriorum nostrorum ibm.

Always provided that the said Thomas, Robert, Richard, and William (the representative burgesses) shall make and repair the said mills at their own costs, taking sufficient structural timber for such making and repairing as may be required from our parks of Toxteth and Croxteth and our forest of Simonswood, by view and liberation of our foresters there.

In 1399, the Duchy of Lancaster being united to the Crown by the accession of Henry IV., leases of town or mills were granted or renewed by the sovereign; and though in various renewals "our mills" are included, they are not specified. In 1410 another allusion to lands juxta Euerstan milne occurs.

Add. MSS., 32105. 469.

Katharine, widow of William Fil Adam, married, secondly, Thomas del Crosse del Hough, of Wigan; and on his death settled her interest in the mill, in 1412, upon her son by her first marriage, John de Liverpool, junr.; but eventually it passed to Richard Crosse, her son by her second marriage, in whose family it long remained:—

Crosse Deeds, 120.

Sciant p'sentes & ffut'i q ego Kat'ina del hogh quondam vxor Willī ffilij Adē de Lyv'pull in pura viduetate mea dedi ĝsessi & hac sursū reddedi Ioh'i de Lyv'pull juniori ffilio meo totu statū meū q ħeo seu quonism' habere pot'o in vno Molendino ventritico cū trib3 selionib3 adiacentib3 in villa de Lyv'pull ac eciam in una placea t're vocata Le Tauerne iacente ī villa de Lyv'pull in Le Bonkstrete videlicet int' tenement Ricī de Stanay ex una parte et tenement scī Nycholaij ex altra parte. Habend et tenend p'dtm̄ molendinū cū t'b3 selionib3 ac coīa p'dtam placea cū oib3 suis ptinēcis p'fato Ioh'i heredib3 et assignat' suis de capitalib3 dūis feodi illius p seruicia inde debita et de jure ĝsueta inp'petm̄. Et ego vero p'dta Kat'ina & her' mei p'fata Ioh'i hereb3 et assignatis suis ĝtra oiēs gentes warranti3 et inp'petm̄ defend'. In cui rei testimoniū huic p'senti sigillū meū apposui hiis testib3—Roberto de Derby, tunc Maiore ville de Lyv'pull, Ioh'ē Osboldston, Nycholas le Clerk, Ioh'ē del Hogh, et alijs.

Dat apud Lyv'pull die Lune proxi post ffestū Annunciacōis beate Marie Virginis anno regni regis henr quarti t'tio decimo.

Be it known to all, present and future, that I, Katharine del Hogh, formerly the wife of William the son of Adam of Liverpool, have, in my widowhood, given, conceded, and by this present confirmed to John of Liverpool, junr., my son, all right that I have or ever could have had in one windmill with three selions adjacent in the town of Liverpool; also in one plot of land called The Taverne, lying in the town of Liverpool in Le Bonk Street, between the tenement of Richard de Stanay on the one side and the tenement of St. Nicholas on the other. To have and to hold the said mill with three selions, together with the said plot with all appurtenances, by the said John, his heirs and assigns, under the chief lord of that fee, by services due and customary therefor in perpetuity. And I, the said Katherine, and my heirs shall warrant and defend the same to the said John, his heirs and assigns, in perpetuity. In testimony of which I have placed my seal hereto in the presence of Robert de Derby, now Mayor of Liverpool, John de Osboldston, Nicholas the priest, John del Hogh, and others. 3. Eastham Windmill. Leases, 1257–1413.

Given at Liverpool on the Monday next after the Feast of the Annunciation of the Blessed Mary the Virgin [March 25], in the thirteenth year of the reign of King Henry the Fourth [1412].

Sciant quod ego Ricus del Crosse dedi Katherine del Hough unum molendinum ventriticum cum tribus selionibus adiacem in villa de Lverpull cum omnibus suis pertinentijs. Habend prefect Katherine ad totam vitam sua de capitalibus dominis feodi illius per seruic inde debita consueta. Test Robert de derby maiore &c. Add. MSS., 32105. 498.

Dat apud Liv'pull du Sabbi proximū ante festum Sancti Ambrosii anno 13 H. 4.

Be it known that I, Richard del Crosse, have given to Katharine del Hough one windmill and three selions of land adjacent in the town of Liverpool, with all appurtenances. To have and to hold by the said Katharine for her entire life, by the usual services due therefor to the chief lord of that fee. Witnesses: Robert de Derby, Mayor, &c.

Given at Liverpool on the Saturday next before the Feast of St. Ambrosius [April 4], thirteenth year of Henry IV. [1412].

Sciant quod ego Katherina quondam uxor Thome del Hough dedi Rico del Crosse heredibus et assignatis suis unum molendinum ventriticum cum tribus selionibus adiacentibus et omnibus suis pertinentijs in villa de Liverpoole. Habendum et tenendum predictum molendinum cum tribus selionibus et pertinentijs predicto Rico heredibus et assignatis suis ad totam vitam. Reddendo inde annuatim mihi viginti sex solidos et octo denarios. Ibid., 32105. 452.

Dat apud Liverpoole die luna proximū post festum Sancti Bartholam anno 13 H. 4.

IV. KING'S MILLS, LIVERPOOL.

3. Eastham Windmill. Leases, 1257–1413.

Be it known that I, Katharine, formerly the wife of Thomas del Hough, have given to Richard del Crosse, his heirs and assigns, one windmill with three selions adjacent, and all appurtenances, in the town of Liverpool. To have and to hold the said mill and three selions and appurtenances to the said Richard, his heirs and assigns; rendering therefor annually to me twenty-six shillings and eightpence.

Given at Liverpool on the Monday next after the Feast of St. Bartholomew [August 24], thirteenth year of Henry IV. [1412].

4. Eastham Windmill. Lease and Repairs, 1450.

4. By 1450 the Crosses' holding had terminated; and "our two mills" are again specifically mentioned as leased out singly and apart from the town; Edward Crosse, receiver for the Crown in 1451, thus accounting for their rentals:—

Ministers' Accounts, 29-30 Hy. VI.

Et de xiij^s iiij^d de firma molend. ventritici voc new mylne sic dim Robto More ad t'mi annoꝫ ꝑ indentur penes ipsm reman:

Et de xxvj^s viij^d de firma alt'ius molend ventritici ibm voc Etton Mylne sic dim Ioh'i Fleccher ad t'mi annoꝫ.

13s. 4d. for the farm of the windmill called New Mylne, as devised to Robert More for a term of years by indenture in his possession:

26s. 8d. for the farm of another windmill there [in Liverpool], called Etton Mylne, as demised to John Flecher for a term of years.

The so-called "new mill" was that of Townsend; its rental being only one-half that of Eastham, then and later the largest and most important king's mill in Liverpool. The above compotus proves that it was not a "new mill," like Townsend; and, in fact, Crosse also accounts for payments made for its repair:—

In div'sis custibꝫ ꝑ appartorem ibm̄ hoc anno apposite suꝑ reꝑacoē moli ventritici voc Etton mylne ut in oꝑibꝫ carpeñ p. billam de ꝑcell suꝑ compm̄ examinate iiij^s vij^d.

In divers expenses for workmen incurred there during this year for the repair of the windmill called Etton Mylne as regards carpentry work, per bill of particulars certified in this account, 4s. 7d.

Close Rolls, 11-13 Edw. IV.

In 1471 Edward Crosse himself held the fee-farme lease of the town and its two windmills, and in 1475 Henry Crosse was lessee of the same. The total rental was £14, the amount at which it remained till 1528, when it was increased to £14 6s. 8d. It is by the amount of the fee-farme rent at this period that we may know the two mills were included in the town lease, though they are not specially mentioned in the deeds. IV. KING'S MILLS LIVERPOOL.

5. In the next century occurs almost the only evidence in the town books of the corporation exercising a control of the mills; two enactments being transcribed in the records of 1558 from a book of *Precedents* of about 1541, which is stated to have been then presented to the mayor. 5. Eastham Windmill. Corporation Regulations, 1541.

It is ordered by the Great Inquest [the chief Assembly of the year] that no man shall sell any ground malt within the town, upon pain for every fault xij^{d}; except the millers their toll malt. Port Mote, i. 41.

The practice of evading soke by purchasing flour and malt ready ground at some mill outside the town, beginning to creep into many astricted districts at this period, is found to be prohibited at various places throughout the country, as by regulations of the Millers' Guild at York, and soke restrictions at Wakefield; while at Chester it was suppressed by stringent prosecutions. The other Liverpool regulation of 1541 was:— Text, III. 122, 259; IV., ch. III.

Every miller, upon warning to them given, shall bring their toll dish to master mayor to [be proved of] a lawful size and sealed, upon pain of vj^{d};—

a salutary usage, already illustrated as of very general observance. Ibid., III., ch. v., § 10.

6. It becomes necessary to interpolate here what little history attaches to another mill; for at this stage the long connection of the Crosse family with Eastham lease was about to be broken, and the equally long association of the Molyneuxs of Sefton with it to be commenced. Sir William Molyneux had held the sublease of the town since 1537, and his son, Sir Richard, had obtained from Henry VIII. in 1546 the lease in chief. But in 1553, when the latter was renewed, the Crosses seemed to have purchased, or otherwise gained, 6. Middle Mill, 1557. Duchy Leases, Div. XI., xxxii. 17.

IV. KING'S MILLS, LIVERPOOL.

6. Middle Mill, 1557.

Inq. p. m., 4-5 P. & M.

Duchy Leases, 4-5 P. & M. 34.

the consent of the Crown to their erecting a mill of their own in the town. For when, in 1557, John Crosse died, he was found by inquisition to have been possessed of one windmill in Liverpool held "de rege et rega"; being built, apparently, within the previous four years by consent of the Crown, and vested with a soke right over that part of the town in which his estate lay. This was a sub-division of town multure little approved of by Molyneux; who in the same year, when renewing the general lease of the town to the corporation, expressly "alwayes forpossessed and reseryved all the [royal] milns standing and beying in the said towne or borrowe or the libertys of ye same."

Crosse's new mill was built near the family seat, Crosse Hall * (by the present Crosshall Street, Dale Street), on the slope of the Liverpool Dale (on the site of the present North Street),† a quarter of a mile distant from Eastham Mill on the opposite site of the creek. From its central position in the town it was called Middle Mill ‡; while the dale below it was called Middle Mill Dale in contradistinction to Eastham Mill Dale on the opposite bank, with which in modern times it has, however, been much confused, even by the local antiquary Okill.§

Middle Mill proved a sore thorn in the side of

Moore Deeds, 347. — * 1519. Land in le Dale strete opposite capitalis messuag Rogi Crosse.

Crosse Deeds, 223. — 1742. Crosse Hall bequeathed by Richard Crosse to Ann his wife.

Ibid., 173. — † 1513. A close in Liverpool called Le Middil Hey.

Ibid., 214. — 1697. A field near Crosse Hall belonging to John Crosse, called Middle Field.

Port Mote, ii. 686. — ‡ 1610. Lease of two parcels of land in the town's field of Liverpool, whereof one is called Middle Mill Dale and the other is called Mill Dam [Eastham].

Ibid., iii. 699. — 1659. We order that Mrs. Crosse and Thomas Birch shall make a gate for a cart-pass between Middle Mill and the town's common as formerly.

Ibid., iv. 29. — 1672. The town's common lying in the bottom of the Middle Mill croft—namely, lying east of the Middle Mill Dale. Also land in Middle Mill Dale abutting to the Town's Gate, and so at the bottom of Mr. Crosse's field.

Ibid., x. 601. — 1753. Lease of the Middle Milne Dale and the little meadow at the bottom of Dale Street.

Okill, W. Dby. MSS., iii. 205. — § "On the site of the north side of Circus Street the stream supplied a mill-dam called Eastham *or Middle* Mill."

IV. KING'S MILLS, LIVERPOOL.

Molyneux, when, as lessee of the royal soke, then chiefly centred at Eastham, he was compelled in 1587 to bring the extensive action shortly to be recounted, against a compact combination of burgesses for absenting themselves from the royal mills. It was admitted on this occasion by the recalcitrant defendants that "they have had their corn ground at the winde milne of one John Crosse, esquier," who was evidently, therefore, to be reckoned among the antagonists of Molyneux. However, Middle Mill kept its district soke intact for some time. Here it was that, apparently about 1635, originated in Liverpool the custom of carrying wheat and flour to and from the mill by the miller; the novelty evoking the displeasure of Moore, of Townsend Mill. In 1641 another John Crosse was found by inquisition to have been possessed (as of a mesne manor) of lands, messuages, a dove-cot, and a mill "held of the king as of his Duchy of Lancaster in free burgage by 23s. 10d. yearly rent, and worth by the feodary's certificate 80s." In 1672 the tenant of Middle Mill was one of several local millers fined for grinding on Sunday. Little record remains of the later history of the mill, which lasted till the days of soke were over, and is marked in Enfield and Perry's map of Liverpool for the year 1768.

6. Middle Mill, 1557.

Text, post, p. 167.

Text, post, p. 180.

7. The year 1557 was a period of conflict between Molyneux, the royal lessee, and his tenants, the corporation and burgesses. A dispute which arose as to the right of collecting market and other toll spread to the soke of the mills, and Molyneux appealed to Queen Mary. On June 4th, 1557, the following order to enforce the soke was directed to the sheriff:—

7. Eastham Windmill. A Royal Command, 1557.

A Commandement that the tenants of and within the maner and towne of Liverpole shall grynde there corne and grayne at the Kinge and Queenes Mills ther and not ells where.

Duchy Commissions and Orders, Edw. VI. —Elizabeth 199.

IV. KING'S MILLS, LIVERPOOL.

7. Eastham Windmill. A Royal Command, 1557.

Whereas we and o^r^ predecessors of us the Quene have had of ancient time divers and sundrie milles within our maner of Westdarby and alsoe in our towne of Liverpole, and have had yerlie rents and pffits for the same the time owte of memorie of man: and that o^r^ tenantes and Inhabitance within o^r^ seid maner of Westdarby and o^r^ seid towne of Liverpole have used, the time owte of memorie of man, to grende there corne and grayne at the seid milles, whereby our seid milles have byn mayntened, our Rentes truelie aunswerid to us and o^r^ predecessors, for us the Quene and o^r^ fermers and occupiers, some reasonable gaynes:

And as we byn creaditlye enformed that there be divers of o^r^ tenauntes and other Inhabitanse of o^r^ seid towne and o^r^ maner of Westdarby, where you be o^r^ stewarde, that have nowe of late erected, builded and made divers and sundrie milles of divers kende of sortes with^n^ o^r^ seid towne as alsoe ells where with^n^ o^r^ seid maner of Westdarby in our seid county of Lancaster: and that there byn divers and manie of o^r^ seid tenauntes and alsoe divers other psons, wch have byn used and accustomed to griende there corne and grayne at o^r^ seid milles, that now of late have grended ther seid corne at the seid milles so latlie builded; whiche hathe not onlie byn to the grate hurte and damage of o^r^ fermers ther, but alsoe with^n^ shorte time to cū may growe to the utter decaye of our seid milles ther, and alsoe to the disinherison of us and our herers and successors of us the Quene:

Wherefore we, mindinge the Reformacō̃n thereof and the pservacon of owre inheritaunse in the p'misses, will and requier you and never the lesse charge and comaunde youe that youe immedeatlie upon the sighte hereof, by v'tue of thes o^r^ lrẽs, doe geve monicō̃n and warninge: And also doe charge and comaunde all and ev'y o^r^ tenauntes, fermers and copieholders dwellinge wth^n^ o^r^ seid manor and towne and either of them, that theye from the time of the seid monicō̃n and warninge doe greinde all there corne and grayne at somme of o^r^ said milles and in noe other place: And alsoe that youe doe geve like monicō̃n and warninge to all other persons dwellinge and inhabitinge wth^n^ the seid maner or towne that they and ev'y of them wch have byn used and accustomed to greinde ther corne and grayne at auncient milles wth^n^ the seid manor or towne shall from thenseforthe greynde ther corne agayne at the seid aunciente milles, and not at anie mille erected and builded in the seid maner or towne wth^n^ the time of xx yeres last past, other then suche as be of o^r^ milles, yf anie suche be; unto suche time as they shall shewe good matter before o^r^ Chancello^r^ and Councell of o^r^ seid duchy whie and wherfore they owght not soe to doe:

And that youe by v'tue herof doe geve comaundemt and charge to ev'y of the seid psons that shall doe to the contrarie of entente and effecte of thes o^r^ lrẽs that they and ev'y of them shall appere at a certen daye by youe to be lymited in o^r^ duchy Chamber at Westm̃ then and ther to aunswere ther facts and doinges therein: And that youe in like maner send to us unto o^r^ court of o^r^ seid duchy the note of such facts and doeings whiche they have comitted and done cō̃trarie to the effecte of thes o^r^ lrẽs:

7. Eastham Windmill. A Royal Command, 1557.

And that youe from time to time doe geve in charge to suche inquests, and shall trulie enquier and true psentment make, of all o^r^ seid tenauntes fermors and copieholders of o^r^ seid maner and towne or other psons dwellinge or inhabitynge with^n^ the seid maner or towne whiche have byn used and accustomed to grinde ther corn and grayne at o^r^ seid milles and goe frome o^r^ seid milles to griende at anie other mills then o^r^; and putte ther names and ther facts and doeings therin, and wher they micht have bin and hadd ther corne grened at o^r^ seid milles :

And upon youre discrecõn to see suche as shall offende in this behalfe be fined and amersed for the same: and that the same shalbe streted and gethered to o^r^ use wth ther fines and amercem̃ts owte in the saide courte : And yf they that shall soe be ps̃nted in o^r^ seid courte upon monicõn to be geven to them by youe or yor deputie will not doe ther duties to o^r^ seid milles, without resonable cause to youe or o^r^ stewarde there for the time beinge showed and the same by youe or other o^r^ stewarde ther to be admytted and alowed, and ther byn licens to them geven for a time ; but obstantlie doe goe and w^th^drawe them from o^r^ seid milles, that then youe or other o^r^ stewarde ther shall geve them and ev'y of them, by v'tue herof, comaunde to appere before o^r^ Chauncello^r^ and Councell of o^r^ seid Duchy of Lancaster in o^r^ Duchy Chamber at Westm̃ at a certen daye to be lymitted, then and ther to aunswere to the p'misses to the entent that o^r^ seid Chauncellor and Councell of o^r^ seid Duchy maye see o^r^ inheritaunse in the seid milles p'servid, and the seid offendars punished ; And further to proceed therin accordinge to the order of our lawes and monicõns and justice : accordingelie not faylinge, as youe will answere to us at yor ꝑell.

Geven the fourthe daye of June in the iiij^th^ and v^th^ yeres of o^r^ Reignes.

This monition and warning, severe as it may appear, was yet, for the times and under the circumstances, an order of comparative leniency. It was issued by Mary, not as a sovereign, but as the lady of the manor. By the soke of the town "our said mills have been maintained and our rents truly answered"; its loss "may within short time to come grow, to the utter decay of our said mills and also to the disinherison of us and our heirs"; and it is desirable that the Chancellor "may see offenders punished and our inheritance preserved." It appears to have further embittered the hostility constantly manifested against Molyneux by the town, between

7. Eastham Windmill. A Royal Command. 1557.

whom little love was ever lost;* but it certainly suppressed the rebellious tendencies of the burgesses for another thirty years, during which Molyneux, holding a lease in chief of all the mills, let them to different speculators or millers. Eastham in 1557 may have been already tenanted by the Boltons, millers, who are found there shortly after; but Townsend in the same year was actually in the hands of speculators who let it to a tailor.

8. Eastham Windmill. Evidences, 1563–1606.

8. Various incidental allusions to Eastham Mill continuously demonstrate its existence; and one in 1563 indicates with tolerable distinctness its site, by mention of "the lane that driveth or leadeth from St. Patrick's Cross into the town field towards Eastham Mylne"; the cross being near the top of the present Hatton Garden, and the lane being the present Great Crosshall Street. In 1566, when to Queen Elizabeth plaintively appealed "your poor subject Ralph Sekerston [M.P.] of your grace's decayed town of Liverpool," desiring that she might "relieve us like a mother," he duly reminded the queen of the royal estate in "your majesty's own town," including its "two windmills"; one of which was Eastham.†

Port Mote, i. 149.

Ibid., i. 370.

Ibid., i. 183.

* 1563. This year war was proclaimed in France and nowhere else. My lord the earl, my old lord of Derby, gave the town a bucke, a pure good one, and merellie disposed of and eaten in the common hall. Also Sir Richard Molyneux gave the town a bucke which proved but mean; it was divided between Mr. Mayor, the aldermen his brethren, and the bailiffs; whereat many commoners lowered and something murmured.

† The corporation did not seem to be specially gratified with Sekerston's achievement in appealing to the lady of the manor on their behalf. It is, however, but justice to the city fathers to note that they never forgot that "our sovereign lady the queen" was lady of the manor whose affairs they conducted. As was due, it was in her name that the port-mote Cura was convened in the Aula Burgi; there the mayor presided as "mayor for the said lady the queen"; local inquisitions were held "on behalf of the said lady the queen according to ancient and laudable custom"; and jurors were summoned on behalf of the Crown:—

Ibid., i. 240. 1565. Inquisit. capt. apud Portmote coram mrō Johne Crosse p' antedcā Dnā Reginā maiore.

Ibid., ii. 131. 1576. Convocatiō in aula coram mrō Thoma Bavand p' dcā Dnā Rnā maior.

Ibid., ii. 272. 1583. Curia portmot magnū tent vicesimo primo die Octobris anno regni dnē Elizabethe regina &c. vicesimo quinto coram Rado. Burscough gen, pro dcā Dnā Regina maiore. Inquisitiō capt ibm. pro dcā Dnā Regina scdm̄ anticū et laudabil consuetudinem in aula coī ville p'dcte.

Ibid., ii. 854. 1619. Noiā jur. ad inquirend p. Dnō Rege.

IV. KING'S MILLS, LIVERPOOL.

8. Eastham Windmill. Evidences, 1563–1606.

In July 1577 Port Mote "enquired of the casting down of a stone coppe or ditch, being near unto Estham Mylne, adjoining the land of Mr. Starkie, which was cast down by the appointment of Alice Abraham, widow, late wife of Richard Abraham"; for which offence she was fined in October following twelve pence.* In 1586 Queen Elizabeth renewed Molyneux's lease; he to "repair, sustain, make and maintain the aforesaid mills, as well concerning stones called millstones as otherwise, at his own proper costs and charges." In 1587 public hostility to the mills was once more rampant, and on this occasion Molyneux had recourse to law; bringing an action against the prominent offenders, as related more appropriately in connection with Townsend Mill; and entering on his bill of complaint Eastham Milne first in the list of holdings under the town lease. In the next year (1588) evidence of identification of the site occurs; Richard Gelybrand granting Ralph Sekerston certain "land in the town field of Liverpool, in a place called Estham Dale, viz. the fifth land from the wyndmille called the Estham Milne." In 1589 the Molyneux rental † includes the receipt of 40s. per annum from John Bolton, of Eastham, for "the mill there called Eastham Milne"; out of which sum 25s. was payable to the Crown. In 1598 Richard Bolton is found paying the same rental; and again in 1606.

Port Mote, ii. 144, 153.

Text, post, p. 160.

Okill MSS., xiv. 28.

9. Corporation Intrigues, 1629–1644.

9. But the ravages of civil war and the abrupt close of the history of the great windmill at Eastham were at hand. In 1629 Charles I., pressed for funds, sold to the Corporation of London his inheritance

* Starkie's land was on the heath at Eastham. Richard Abraham, above mentioned, was tenant of this plot and of the neighbouring Gallows Field in 1562; and in 1610 Isaac Abraham, yeoman, held "Milne Hey common in the town field of Liverpool, lately in the occupation of Alice Abraham, his mother." Port Mote, i. 149; Ibid., ii. 688.

† Consulted at Croxteth Hall, by the courtesy of the Earl of Sefton.

IV. KING'S MILLS, LIVERPOOL.

9. Corporation Intrigues, 1629–1644.

of the manor of Liverpool. In the deed of conveyance the mills were not specifically mentioned; but that they were included in the sale is evident from the total annual reserved rent to be paid to the Crown—£14 6s. 8d.—the amount at which it had remained since the time of Henry VIII. In the next year occurs the last allusion to Eastham Mill as still standing; the corporation directing in 1630 that "Richard Rose of Everton shall remove the stones that now lie upon the usual way from Estham Milne, and that the way shall be where it was anciently afore." In 1635 Molyneux, seizing his opportunity, purchased the fee-farme of the town from the speculators who had acquired the same from the London Corporation, thus becoming actual lord of the manor. The deed contains the usual proviso—"together with all and singular farms, messuages, cottages, *mills*, &c., situate within the said town"—which, though summarised by Picton as "additional verbiage, mere words without any specific application," nevertheless definitely conveyed to Lord Molyneux the shortly-to-be-contested ownership of the mills.

Port Mote, iii. 101.

Memorials of Liverpool, 1875, i. 115.

In 1638 the corporation approached Charles I. with the object of ejecting Molyneux. On May 21 of that year it appeared that the mayor and two aldermen had been to London on the business of an action brought against the town by Lord Molyneux; and that they had presented a petition to the king, calmly referring to Molyneux's right of possession by purchase as "the claims of Lord Molyneux in virtue of a late grant made by the Londoners [the Corporation of London, who had purchased Liverpool] to the said Lord Molyneux"; and requesting the king, "if it please his Majesty, to grant the said town and the whole emoluments thereof to the mayor, bailiffs, and burgesses of Liverpool

Port Mote, iii. 228.

9. Corporation Intrigues, 1629–1644.

and their successors for ever in fee-farme." The king, they reported, had referred the petition to the Lord Treasurer and the Chancellor of the Exchequer. On May 24, as "great sums of money will be required to expedite this business at London," a list of subscriptions was opened, the total reaching £162. But if these great sums had been forthcoming to replenish the depleted royal coffers it is not quite clear how, as the king had sold the estate, he was to be expected to deprive Molyneux—a Royalist, moreover—of his purchase. However, nothing came of the scheme, and at about this period the corporation forsook the Royal cause and became adherents of the Parliamentarians.

On March 5, 1644, the anti-Royalist town having been stoutly though ineffectually garrisoned against Prince Rupert, the corporation met to consider "the present state of the town and the losses and sufferings of the inhabitants by reason of the cruelty of the prince's army"; passing on ultimately to a resolution that the Recorder should go to London and interview the authorities on whose behalf they had suffered so severely, with a view of procuring some recompense for the same. No more likely suggestion offered than the spoliation of the Royalist Molyneux; and accordingly their petition included a request that for the better government of the town "the milnes and ferry boats formerly belonging to the town, and injuriously taken and kept from them by the Lord Molyneux, might be restored to the corporation as formerly";—much the same prayer to Cromwell's Parliament that, backed up with "great sums," had been preferred to King Charles six years before. The mainspring in this action was undoubtedly "Collenoll John More, governer of this towne of Liverpool"; whose name, following upon

Port Mote, iii. 366.

IV. KING'S MILLS, LIVERPOOL.

9. Corporation Intrigues, 1629–1644.

that of the mayor in the record, testifies to his presence and influence at a meeting of the corporation in the previous December, at which it was ordered that the goods and merchandise of "rebels, traitors, felons and other malefactors" should be seized by the corporation. He was of the ancient local family, seated at Bank Hall, long associated with Townsend Mill.

The appeal of the Parliamentary corporation to the Cromwellian Legislature proved successful, and in October 1646 an order of Parliament invested the borough with the sequestrated Molyneux estate in the town:—

Port Mote, iii. 397.

1646. Oct. 1.—Whereas a Wyndmillne and ferry boats formerly belonging to the Corporation of Liverpool were late in the possession of Richard Lord Molyneux, who is in hostility against the Parliament, and by his power with the Lord Cottington, late Master of the Courte of Warde and Liverie, brought a vexatious suit against the said Corporation, to their great damage and impoverishment: And whereas all the writings and ancient records belonging to the said Corporation were taken away when that town was taken by the enemy: Considering the exceeding great losses and sufferings of that said town, and to the end that the ancient right of the said Corporation may be restored, &c.; the Lords and Commons assembled in Parliament do order and ordain that the said Corporation shall have, hold and enjoy the said Wyndmilne and ferry boats and the rent of £20 per annum formerly paid by the said Corporation to the said Lord Molyneux, till both Houses make further order.

10. Eastham Windmill. Destruction, 1644.

10. But while in 1644 the corporation had prayed to be granted "windmilnes," the above order only empowered them to seize "a wyndmillne." This was that of Townsend, the only one left on the Molyneux estate in Liverpool; the other, Eastham Mill, having apparently been destroyed during the siege that had closed with the capture of the town by the Royalists, June 2, 1644.

In 1650, the year after the death of Charles I., the corporation lost no time in completing their holding over the town; deciding that "the fee-farme rent of this borough, which is £14 6s. 8d. per annum,

Port Mote, iii. 502.

shall be bought and contracted for with the trustees of Parliament"; that is, that the reserved rent which had been paid by Molyneux to the Crown (after his purchase), and was now payable by the corporation to Parliament (after their grant), should be compounded for and extinguished. This they effected, and the old "King's Majesty's Corporation" became *de facto* "Lords of the Manor of Liverpool" and of all the sources of its manorial revenue, including the remaining mill. In 1657 they made a "Rentally of burgage rents lately purchased from the trustees, according to Act of Parliament"; including in it the entry, "Edward Moore, esq., for the Townes end Milne and the horse milne, and certain parcels of land," but of course making no allusion to the extinct mill at Eastham.

10. Eastham Windmill. Destruction, 1644.

Port Mote, iii. 651.

In 1654, and the year following, the corporation, as lords of the manor, entertained the idea of rebuilding Eastham Mill:—

1654. Dec. 19.—It is ordered by the Assembly that the present mayor and bailiffs shall bargain and issue out monies for the rebuilding of Eastam Milne, and shall take course to preserve the materials that are left and make sale of such as will not serve again: And shall cause Mrs. Bixsteth to deliver to them the old millstones which Mr. John Williamson borrowed and took from thence. Ibid., iii. 605.

1655. Aug. 22.—It is again ordered that Mr. Mayor and the bailiffs shall treat with Richard Whytehead or any other about the setting up of Eastam Milne: and shall have power to contract and agree with him or any other for a certain barn and rent for the best advantage of the Corporation: and to reserve the ancient rent of twenty shillings per annum burgage: and make what more they can well agree for. Ibid., iii. 610.

1655. Nov. 20.—Ordered, that Mrs. Bixsteth shall restore a milnstone which was taken from Eastham Milne or give satisfaction to the bailiffs for it.* Ibid., iii. 622.

The corporation, however, never restored the mill.

* In the rental of 1657 appears "Margaret Bixsteth, widow, for the house late Williamson's, and more for her barn"; and in the rental for 1668 are "Mrs. Bixsteth for a barn" and "Mr. Williamson for John Stursaker's house, and more for the Eastham milner's messuage." In 1695, when soke was practically abolished, one Thomas Bixsteth was granted liberty by the corporation to build a windmill. Ibid., iii. 651. 871. Ibid., iv. 364.

IV. KING'S MILLS, LIVERPOOL.

10. Eastham Windmill. Destruction, 1644.

In 1660 the Restoration changed the face of affairs; and the erstwhile Royalist corporation, that had sought to curry favour with and bribe Charles I., and then, in his waning days, had become Parliamentarian and accepted sequestrated Royalist properties, once more sailed safely with the times and became Royalist; seeking on August 20, 1660, to avoid the frown of Charles II. by a complete surrender of their lately acquired burgage rents:—

Port Mote, iii. 706.

> The fee-farme rent of £14 6s. 8d., lately purchased by this town, shall be granted and resigned to the King's Majesty by a grant under the Town's Seal, with an humble address.

Thus Lord Molyneux eventually came to his own again, but neither did he rebuild Eastham Mill.

Round its ruins the old place-name lingered some little time longer. In 1679 it was ordered that "a sufficient bridle way be set out from the narrow way which leads from Everton to Eastham Mill, according to ancient custom"; and in 1691, for not paying homage to Liverpool Port Mote, a fine was inflicted on one, "James Goore of Eastham."

Ibid., iv. 111, 290.

The approximate site of the structure was on the high ground on the east side of the present Thurlow Street, overlooking the pool of the watermill in the Dingle, in Down Street. Like Townsend Mill, which stood only about a couple of hundred yards away, it was of course a timber erection of the primitive post type.

Text, ante, p. 136.

11. Townsend Windmill, 1347–1450.

11. Townsend Mill does not appear in the records for nearly a century after 1257 when Eastham windmill was in existence. Duke Henry, who died in 1361, had possessed in Liverpool "two windmills and a horse-mill"; one of the two windmills being that of Eastham, the other that of Townsend. In an *Extent* of the duke's possessions in 1347 they had been stated to be worth 10 marks (£6 13s. 4d.) per

annum; and, in the receiver's accounts for 1348, to be leased, together with the farm of the markets, ferry-boats, etc., to John, son of William del More—the same William More who for his redoubtable deeds at Poitiers in 1356 was created a knight-banneret on the field.

11. Townsend Windmill, 1347–1450.

At this point may be cleared away a misapprehension that has long prevailed as to the site of Townsend Mill. This has been assumed to have been on the shore at the north end of the town, where were the Moore Hall and the Bank Hall, and other properties of the Moore family. In fact, however, the mill was situated at the east end of the town at Eastham, and within two hundred yards of Eastham Mill. Following the usual custom which had originally led to this latter having been located beside the old watermill, when a second windmill had been established it had been placed not far distant from the first. We shall find abundant evidence confirming this allocation of Townsend to its true site at literally "the end of the town" at Eastham.*

Text, post, p. 159.

By the year 1450 Townsend Windmill appears to have been rebuilt, and in the compotus of Crosse for that year, already quoted, is designated the "New Mill"; being then demised apart from the town to Robert More for a term of years, according to an indenture in his possession, at a rental of 13s. 4d., or one-half the rental of Eastham Mill.

Ibid., ante, p. 142.

* There is extant a published map, stated to be copied from an old MS. map of Liverpool in 1539, over which have been marked the fortifications of 1644; and on this, "Townsend Mill" is marked on the north shore. But no authority is advanced for this map; and as it depicts the so-called Townsend Mill as a tower mill at a period when such mills were unknown, there need be little hesitation in considering it to be fictitious and of much later date than it professes.

Text, II. 293.

About the year 1792, shortly after the destruction of Townsend Mill, a tall tower-mill was erected on the north shore, which the owner chose to call by the well-known name of the extinct old mill. This modern "Townsend Mill," the cause of the confusion above noted, is still existing amid a labyrinth of streets near the Bramley Moore Dock; its lopped tower being used as a warehouse.

App. to this Chapter.

IV. KING'S MILLS, LIVERPOOL.

12. Townsend Windmill, Lease, 1463.

12. The Mores had already become as firmly attached to Townsend Mill as had the Crosses to Eastham and Middle Mills; and, retaining their interest for quite as long a period, they far outstripped them in enterprise as mill-owners; far exceeding them also, however, in evoking popular hostility to their mill, and in aiding to bring about their own eventual ruin. Robert More's lease of Townsend having expired by 1463, it was then regranted at the old rent to him in conjunction with his partner, William Leyland:—

Roll of Indentures, 1–11 Ed. IV. 37.

Rex tradidit & ad firmam dimisit Willm̄o Leylond & Robto More unum molendinū ventriticum in villa de Lyverpole p' antica in tenura dcī Robti. H'eñd & tenend' dcm̄ molendinum cum om̄ibȝ p'ficiis eidm̄ molendino p'tin p'fat Willm̄o & Robto ac assign̄ suis a festo Scī Michis Archī ultimo p'terito usqȝ ad finem t'mini septem annoȝ extunc p'x sequenciū & plenarie complenđ. Reddendo inde annuatim p'fato Dnō Rege heređ & assign̄ suis tresdecim solid' & quatuor denar sterlingoȝ ad festa Pasche & Scī Michis Archī p' equales porcōes. Et p'dcī Willm̄s & Rob'tus sustentabunt & rep'abunt sumptibȝ suis p'prius & expeñ dcū molendinū in om̄ibȝ p't': et p'dcus Dn̄s Rex inveniet grossum mæremiū p' rep'acōe ejusđm in parcis de Toxstath & Croxtath ac bosco de Symondeswode p'cipiend' p' delib'acōem officiār ibm p' tempe existeñ: et illud in finē dcī tm̄ini sufficient' repar' dimittent. Et si contingat dcm̄ annūi redditū xiij^{s} iiij^{d} aretro fore in parte vel in toto p'dimidiū annū post aliquem t'minū quo solvi debeat non solut, extunc bn̄ licebit p'fato Dnō Regi hered' & assign' suis in dict' fermam reintre & illam in manus suas reassum̄ ac in pristino statū suo possidere p'senti dimissione in aliquo non obstant.

Dat &c. apud Westm xxvj^{to} die Februarii anno &c. secundo. P' consiliū ducat'.

The king has granted and to farm letten to William Leyland and Robert More one windmill in the town of Liverpool formerly in the tenure of the said Robert. To have and to hold the said mill, with all the profits belonging thereto, to the aforesaid William and Robert and their assigns, from the Feast of St. Michael the Archangel last past to the end of a term of seven years thence next following and fully to be completed: rendering therefor annually to the aforesaid lord the king, his heirs and assigns, 13s. 4d. sterling, at the Feasts of Easter and St. Michael the Archangel, by equal portions: And the aforesaid William and Robert shall sustain and repair at their own proper costs and charges the said mill in all things: and the said lord the king shall find great timber for the repair of the same in his parks of Toxteth and Croxteth and his wood of Simonswood at the view and consent of the officials there for the time being: and at the end of the term the said mill shall be left in substantial repair. And if it happen that the aforesaid annual rent of 13s. 4d. either wholly or in part remain in arrear for a half-year after it should be paid, then it shall be lawful for the said lord the king, his heirs and assigns, to re-enter upon the said ferme, and take the same into his own hands, even as he possessed it formerly; the present demise notwithstanding.

IV. KING'S MILLS, LIVERPOOL.

12. Townsend Windmill. Lease, 1463.

Given at Westminster, February 24, in the second year of our reign. By the Council of the Duchy.

This lease would expire in 1470. In the next year Edward Crosse has been shown to be lessee of the town and its two windmills; as was Henry Crosse in 1475; the Mores, apparently, during this time working only their horse-mill near More Hall.

13. Townsend Windmill. Lease, 1557.

13. In 1546 Sir Richard Molyneux, of Sefton, acquired the lease of the town and sublet the windmills. That now under notice he granted to two of his neighbours at Sefton, Thomas Eves and John Hetoller, who subleased it to Thomas Banks, of Liverpool, tailor; by whom it would, of course, be once more let to some practical miller. The following is the sublease granted to Banks in 1557:—

Moore Deeds, 368.

This indenture made ye thridde daye of June in ye thredd & iiij yer of the reign of Philipp and Marie by the grace of God King & Quene of englond spayne ffraunce bothe sycill [both Sicilies] Jerusalem and Irelond defenders of ye ffayth archiduc of Austria duc' of Milaine Burgundy & Brabant Counts of hasburge fflandrie & Tyroll

Betwene Thoms̃ Eves & John hehtoller of Sefton w^th^in the countie of Lanc' yoman of the one ptie And Thoms̃ Banke of Liv'poole w^th^in the said countie taylor of the other ptie

Witnessith y^t^ the said Thoms̃ Eves & John hehtoller for dyv'se good consideracõns theym mouing have demysed granted sett & to farm letten & by these p'nte demysith grantith settith & to ffarme lettith unto ye said Thoms̃ Banke all y^t^ Wyndy Mylne all standing and beine w^th^in the Lordshipp of Lyv^r^pole within the said countye of Lanc comonly called & named the Towne End Mylne w^t^ all & singler the mylstones sale yards multes corne of what kynd or nature so ev' hit be, and all comoditys & apꝑtences unto the said Wyndy Mylne wth thappurtencẽs.

To have hold obtayne & injoy ye said Wyndy Milne & all & singler the ꝑ'misses w^t^ thapp'tencẽs unto ye said Thomas Banke his ex^rs^ and assigns to his and thair vses ꝑfit & advantage imediatly

13. Townsend Windmill. Lease, 1557.

from & after ye day & feast of ye Na^{t} of Saynt John bapt next ensuyng ye dat hereof unto ye complete end & t'me of sextene yers then next ffolowyng fully to be completed and endyt.

The said Thom̃s Banke his ex^{rs} & assignez yelding & payying yerly & ev' yer during the said terme ffor the said Milne & other ꝑmisses ye tythe & Quenes mat' rent due and akustomed at ye fest days there used viz xx^{d} yerly unto ye Vycar of Walton ffor ye tythe berne & the tythe of ye said Mylne.

And ye said Thom̃ Banke for himselfe his ex^{rs} & assigns doth couenñte & graunte to & wth the said Thōs Eves & John hyghtoller thair ex^{rs} & assigns by these p'snts y^{t} he the said Thomas Banks his ex^{rs} & assignez shall yerly & eũe yer during ye said terme well & truly content and paye or cause to be contented & paid unto ye said Thoms Eves & John hyghtoller thair ex^{rs} & assigns ffor ye said Wyndy Mylne & other ye ꝑ'misses ye some of ffaure iiij^{li} of usuall money of England ow' & besyd the tythe & Queenes mats rent & xx^{d} ffor ye tythe as is aforesaid at ye feasts of Saynt Martyn ye bysshoppe in Wynt^{r}, & rent, by evyn porcōns or wthin iij dayes next ensuying eth^{r} of ye said feasts

And more ov' ye said Thomas Banke doth couenñte & graunte to & with ye said Thomas Eves & John hyghtoller by these p'sents y^{t} he ye said Thomas Bancke his ex^{rs} & assignez shall & wyll well & substancially ffrom tyme to tyme during all ye said terme of xvj yers upon his & thair alone ꝑpr costs & charges reparell sustene buyld & kepe up ye said Wyndy Mylne & ev' ꝑt thereof: And do leave ye said Mylne w^{t} hit app'tẽnces at ye end of ye said terme well & substancyally repairelled & upholden in all things:

Ꝑvided alwayes y^{t} ye said Thomas Banke his ex^{rs} and assignez shall not be charget to ffynde & delyv' at ye end of ye said terme sale clothes, pycks [millstone picks] ne Crotte of Corne . . . . as ye same are ye ꝑpr goods of ye said Thomas Banke enything conteyned in these p'sents to the contre not w^{t} standing:

And the said Thomas Eves and John hyghtoller doth couñte & graunte to & w^{t} ye said Thomas Banke his ex^{rs} & assigns by these p'sents y^{t} if all & singler ye couñts graunts & artycles covenanted & limitted upon ye ꝑtye & behalf of ye said Thomas Banke his ex^{rs} & assigns conteyned & specyfyed in these indent^{rs} be well & truly ꝑformed accomplished & fulfylled, that then le obligacōn berring det w^{t} these p'sents wherein ye said Thomas Banke Evan haghton gent & Thomas Mylnes joyntly stand bounden unto ye said Thōmas Eves John hytoller their ex^{rs} & assignez in ye some of xl^{li} st'ling, shalbe utt'ly voyd & of none effect Els stande hit in full strength and vertue.

In Wytnez wherof ye p'ties aforesaid to these indent^{rs} int'chaungeably have putte thair seals. Geven ye daye & yere fyrst above written.

Signu manuale Thome Banke.

P'sent at ye enseallyng & delyv'e
herof & witness of ye same Robt ffazakerley
Ihon Neyll Robt Bolton Ryc Tyrer & others.

IV. KING'S MILLS, LIVERPOOL.

14. Townsend Windmill. Evidences of site.

14. As the designation—"commonly called and named the Town End Milne"—occurs for the first time in the foregoing deed, some evidences may be adduced here showing that "the Town's End" and the Townsend Mill were at Eastham, and not at the north shore, as already mentioned. Text, ante, p. 155.

In 1562, five years after the date of the above lease, Townsend Mill is shown to be at Eastham in a Rental of Liverpool for that year: "Thomas Secum, now mayor, tenant in fee-farme for ever of the Rough Heys [rough fields, heathlands], lying north-east the Townes End Milne upon the one part, the gallows field upon the west part, Liverpool heath upon the south part, and the lands of Richard Starkie, now in the occupation of Richard Abraham, upon the north part"; the whole of these allusions being already shown to apply to Eastham. In 1566 the Earl of Derby, with a retinue of Lancashire gentlemen, arriving in Liverpool upon a secret commission, was met by the mayor and corporation "at the Townes Ende, between the stone bridge there and Mr. Mayor's hall"; John Crosse being mayor. In 1568 the corporation order the making of the road "from the cross-bridge called Townes Ende Bridge all along through Dale Street to the market-place." In 1575 one of the burgesses is fined for "breaking the Queenes Majesties soyle ground within the liberties of the town, in the highway going down the bank [William Brown Street] from the Mylne upon the Townes End." In 1586 Molyneux grants to William More for sixteen years a lease of "the Winde Mill at Leverpole Townes End"; in 1593 an allusion is made to "the watercourse at the Townes Ende"; and in 1636 was ordered the repair of the pavement between "the Townes Ende Bridge and the Townes End Milne [William Brown Street]." In 1672 the

Port Mote, i. 149.

Text, ante, pp. 134, 149.

Port Mote, i. 322.

Ibid., i. 396.

Moore Papers, Inventory.

Port Mote, ii. 443.

Ibid., iii. 191.

IV. KING'S MILLS, LIVERPOOL.

14. Townsend Windmill. Evidences of site.

Port Mote, iv. 29. Ibid., vi. 428. Ibid., ix. 259.

Rental of Liverpool mentions "the Gorsie Field above the windmill at the Towns End, lying on the west side of Liverpool Heath"; also "the Gallow Field lying to the north of the nearer Towns End Mill," this being Townsend Mill, then standing near the ruins of Eastham Mill. In 1719, to avoid danger of fire, the bakers were ordered to stack their gorse for fuel on the gorse-field, near Townsend Mill, and fetch it thence every day as they needed it. In 1731 James Shaw was granted land "beside the highway [Islington] that leads from Townsend Mill to [West] Derby Chapel." Other identifications of site in 1724 and 1780 are quoted in due course.

15. Townsend Windmill. Action against Burgesses, 1587.

15. The order of Queen Mary enforcing the soke of the royal mills in 1557 lost its force in the town within comparatively few years, and by the year 1587 disaffection had long been rife among the burgesses; William More, as the licensee of Townsend under Molyneux, finding himself a considerable sufferer. More, who had married Dame Eleanor Molyneux, was closely associated with Sir Richard in the legal proceedings which ensued; and the enmity which the corporation and townspeople seem to have consistently exhibited towards the Molyneuxs, practically their landlords, was now vehemently exerted against the unfortunate mill-lessee. He was at this period not only interested in Townsend Mill and the horse-mill at Liverpool, but also in those of Bootle *

Moore Deeds, 624. * N.D. Robert de Byron grants to Matilda his daughter his lands in Botel, with one-sixth part of the watermill there.

Ibid., 625. 1357. Matilda conveys the same to Richard, son of Adam de Aynsargh, of Liverpool; and, on the same day, Robert de Byron, who calls himself "lord of the sixth part of the manor and town of Botell," conveys the same, with the one-sixth share of the mill, to the same Richard de Aynsargh.

Ibid., 270. 1394. Richard Munn, capellanus. *Inq. p. mor.* Was seised of certain lands and the sixth part of a watermill at Bootle; all of which he held of Richard de Walton in right of his church at Walton, paying therefor twelvepence per annum.

Ibid., 355. 1537. Richard Frodsham, capellanus, of Liverpool, grants to William More and Alice his wife lands in Bootle, together with the sixth part of the watermill in Botehill aforesaid, now in the tenure of John Bury at an ancient rent of 11s. 1d.

IV. KING'S MILLS, LIVERPOOL.

15. Townsend Windmill. Action against Burgesses, 1587.

and Kirkdale;* and appears to have set great store on the possibilities to be evoked by a judicious and energetic development of the milling resources of the district. The burgesses, headed by two future mayors, Giles Brooke and John Bird, had deserted Eastham and Townsend; grinding at Crosse's Middle Mill and illicit horse-mills and water-mills within the town, and at distant rural mills, even also at one across the Mersey at Birkenhead. The revolt was widespread and complete, though its justice or injustice as against the millers is not now ascertainable. Certainly the burgesses, in their subsequent capacity of defendants, made out a serious grievance

1544–51. Rentals of Alice, widow of William More. John Bury, for Milne rent, 5s. 6½d.; Henry Bury, for Milne rent, 5s. 6½d. per annum. Moore Deeds, 93.

1567. John Moore purchases the lordship of Bootle. Ibid., 633.

Bootle watermill and the later windmill erected near it remained with the Moores till the dispersal of their estates in the eighteenth century. In 1710 Sir Cleave Moore endeavoured to turn the stream of the decaying watermill to account by obtaining an Act of Parliament to supply Liverpool with water from that source. A copy of the Act is transcribed in Okill MSS., xiv. 90. App. to this Chapter.

* N.D. (*c.* 1290). Le Milnefeld in Kyrkdale. Moore Deeds, 504.

1317. Robert (Dominus de Kyrkdale 1317, *Deed* 523) mentions a road leading from le Breck usq' molend de Kyrkdale. Ibid., 520.

1340. Alice, widow of Robert of Kirkedal, grants away the windmill:— Ibid., 539.

Pateat univ'sis p' p'sentes me Aliciam que fuit vx Robti de Kirkedal dimississe concessisse & ad t'mū vite mee tradidisse Robto de Irlond mol'nd ventriticū quod habui ad t'mū vite mee. Habend & tenend d'to Robto & hered libe q'ete b'n & in pace: Reddendo inde annuatim m' ad tota vitam mea unū qr'tr avena ad f'm annuciacōis bte Marie v'gis & ad f'm Sti Iohis Bapt p' equales porciones p' omibz. Et ego p'dta Alicia & hered mei p'dtm Molend' vent'ticū cū p'tin contra omnes gentes warantizabim^{s} & imp'petuū defendem^{s}. In cuī reī testimoniū huic p'senti sc'pto sigillū meū apposui. Hiis testibz, Simon de Walton, Willo de Stonburgelegh, Ad fil' heyne, Masse de Kirkedall', John de Werall, & aliis.

Dat apud Kirkedal die d'nica prox post f'm Sti Luce evangelist anno regni reg Edward t'cij a conquestū quarto decimo.

Be it known to all. By these presents, I, Alice, who was the wife of Robert de Kirkdale, have demised, conceded, and granted for the term of my life to Robert de Ireland the windmill that I have held as for the term of my life. To have and to hold to the said Robert and his heirs freely, quietly, and in peace; rendering me therefor annually one quarter of oats at the Feasts of the Annunciation and of St. John the Baptist in equal portions. And I, the said Alice, and my heirs shall warrant the same against all persons in perpetuity. In testimony of which I have affixed my seal to this present script. In presence of, &c.

Given at Kirkdale, the Sunday next after the Feast of St. Luke the Evangelist, in the fourteenth year of King Edward III. after the Conquest.

"Thomas del More in ye 19 yere of R. 2. anno 1396 held ye manor of Kerkdell." The lordship remained in the family till the early part of the eighteenth century. Moore Papers, [53].

IV. KING'S MILLS, LIVERPOOL.

15. Townsend Windmill. Action against Burgesses, 1587.

against the mills; but it is to be remembered, that had there been any real gravamen in their complaint—had there been entertained, not a desire to ruin the mills, but simply a wish to remove some incidental grievance—then an appeal to the Duchy Court was quite open to them; and an action, as extensive and concerted as that they adopted to promote rebellion, could just as easily have been adopted to secure legal redress. More took an active part in promoting an action against the offenders, and the initiatory bill of complaint, annotated by him on behalf of Molyneux, still exists:—

Cal. Duc. Lanc., 30 Eliz. iii. 223.

To the Right honorable S^r^ ffancs Walsingham K^t^ principall secretary to her Mat^ie^, and Chancellor of the Duchie of Lancaster:

In most humble wyse complayninge sheweth unto your honor Sr Richard Molyneux of Sefton in the countye of Lancaster Knight That whereas her Mat^ie^ is Rightfully seised in her demesnes of ffee as in the Right of her Duchie of Lanc: of and in fyve Wyndmylnes wth^n^ her Mt^ies^ Manor of West Derby comonly called Eastham Milne Townsend Milne Derby Milne Ackers Milne and Wartre Milne in the s^d^ county of Lanc: nere adioyninge to the townes of West darby and Lyverpoole in the said county: and being so therof seised hath demysed and lett the same to fferme unto your orator by Indenture under the Seale of this Court for divers yeres yet to come yelding therefor yerely duringe the said terme a certen yerely rent and fferme to her Mat^ie^ her heires and successors: And wheras all the inhabitants of West darby and Lyverpoole aforesaide by all the tyme wherof the memory of man is not to the contrary have used and of Right ought to grynd all ther corne at her Mat^ies^ mylnes payinge and allowinge suche Tolle as of auncient tyme hath bene used and accustomed: So it is if it may pleas your honor that one Giles Brookes John Bird Ryc Shawe and divers others inhabitants of Lyverpoole aforesaid have erected and made divers horse mylnes, hand milnes and quernes in Lyverpoole aforesaid and in West Derby aforesaid, and grindeth and do dayly and from tyme to tyme grind at the same very muche mault rye barly otes and other corne and grayne spent in the sev'rall mesuages of the said Town of Lyverpoole and the Manor of West Derby; by reson wherof the custume of grist of her Mat^ies^ said mylnes is very greatly hindered decayed and ympayred, to her Mat^ies^ disinheritance and yor orators hindrance in his saide fferme: So as the p'fit of the said Wyndmylnes . . . [is not sufficient] to pay the yerely rent and fferme reserved and due for the same to her Mat^ie^:

In consideracōn wherof may it pleas your honor to graunt unto your orator the Queenes Mat^ies^ most gracious Writt of Iniunccōn

IV. KING'S MILLS, LIVERPOOL.

15. Townsend Windmill. Action against Burgesses, 1587.

comandinge the saide Giles Brookes John Bird and Ryc Shawe and all others that have erected and made eny the afforesaid milnes or do use or occupie eny of the saide mylnes in Lyverpoole and West derby aforesaid and at any tyme sithence the fferst yere of her Mat^{ies} reign, upon such penalty as your honor shall thinke mete, to surcease and forbeare to grind any corne or malt at the same hereafter, or els to appere before your honor in the Duchie Chamber at Westm^{r} at such day as your honor shall therin lymyte and appoint to shew cause whye they ought not so to do: and also to answere their unlawfull doinge in this behalf: And your orator shall dayly pray &c.

[The Iniunccon must but extende y^{t} selfe to all suche mylnes as be made sence the fyrst yere of her Mat^{ies} reigne and not eny further.]

[Some howes youe most mend the byll wch is putt into the Duche accordinge to this copye and then the Iniunccōn must be made accordynge to the byll; and then all is well.—Wm MORE.]

Moore Papers, 5.

In due course the defence was entered as follows :—

The iointe and severall answeres of John Birde and Giles Brooke defts to the Bill of Complaint of Sr Richarde Molinex K^{nt} complt.

The said Defts saie and either of them for himself saithe That the said bill of Complainte exhibited againste theis defts into this honorable Courte is very uncerten and insufficient in the Lawe to be Answered, for dyvers and sundrie manifest causes and faltes therein appearinge (as theis defts by ther Counselle Lerned are Informd). And the matters therin conteined are sett forthe, continued and devysed as theis defts verelie thinke by one Willm More of Banckhall nere Leverpoole in the said countie of Lancaster esquier in the name of the said Sr Richard Molinex and upon his p'per costes and chardges; ꝓsecuted either upon some private malice conveyed by the saide Willm More againste theis defts or otherwise for his owne lucre and gaine, wthout respect either of her Mat^{ies} benefit or service, or the good estate of the whole inhabitants of the saide town of Leverpoole, beinge an Alderman of the said towne: Neverthelesse the benefitt and advantage of exception to the Incertentie and Insufficience therof to theis defts and either of them at all tymes hereafter saved and reserved—

For answere and plaine declaracon of the trueth to the saide surmysed matters the saide defts and either of them saie that they and either of them are inhabitinge and dwellinge in the saide towne of Leverpoole in the saide countie of Lancaster, wch towne of Leverpoole is an Auncient Burroughe and porte or haven Towne nere adioyninge to the sea coaste, into wch towne aswell divers honorable p'sonages and other her Mat^{ies} officers have used divers times to repaire and resorte for the affaires of her highness s'vices both out and into Ireland. And that her Mat^{ies} forces and soldiers for those pts appointed and ells where upon her highness business have sundrie times usualy taken shipping and landed

IV. KING'S MILLS, LIVERPOOL.

15. Townsend Windmill. Action against Burgesses, 1587.

at the said towne of Leverpoole, sometimes to the nomber of five hundredth sixe hundredth or more: and oftentimes lie there a fourth night or three weeke or longer for lacke of conveniente wynde to transporte them. And theis defts saie that great nombers of people cominge from manie ptes of the countries adioyninge to the saide Towne at the time of the yeare of the herringe fisheinge at suche time as the same doth happen have used for the moste pte to lie and make their abode in the said towne of Leverpoole:

For the better entertainment of which nombres of people and for better p'vysion to be made as well for the inhabitantes of the saide towne of Leverpoole as of all strangers contynuallie resortinge thereunto, there hath bene contynuallie maintained had and used by all the time wherof the memorie of man is not to the contrary (as theis defts verelie thinke) divers horse milnes wth^{n} the said towne of Leverpoole to grinde malt, which have bene erected and sett up as occasion served for the better servinge of the saide towne at their pleasurs.

And theis defts further say that her Mt^{ie} hathe not any water milnes wth^{n} the said towne of Leverpoole and the libties therof nor any other milnes there but onelie twoe winde mylnes whiche are reputed to be her mylnes; and are not sufficiente to serve the whole towne, to the knowledge of theis defts, by reason wherof the Inhabitantes in time of calme wether and at others times have been enforced and constreyned to grinde their harde corne at water mylnes iiij. v. or vj. miles distant: And yet neverthelees theis defts saie to their knowledge that divers and sundrie tymes when, as the winde did serve bothe, as well theis defts as other the inhabitants of the saide towne of Leverpoole have at their Libtie grounde their saide corne at other milnes without beinge compelled to come to her Mat^{ies} said supposed winde milnes; and some time have had their saide corne grounde at the winde milne of one John Crosse esquier,* and Mr. Norres his water milne,† birket milne‡ and other water milnes there aboute at their wills and pleasures where they coulde best be served:

And the saide John Bird one of the nowe defts for himself saithe that he hathe and as ocasion s'veth occupieth and useth one horse milne wth^{n} the saide towne whiche was erected and sett uppe Longe before he was tennente thereunto, being the inheritance of Edwd Norres esquier, as lawfull for him is to doe, as he verelie

* Crosse's "Middle Mill."

App. to this Chapter.

† A private and apparently illicit mill on the upper reach of the ancient Dingle stream. "The watermill at Gallows Mill [London Road] was turned by the stream from Moss Lake Fields" (Okell MSS.). In 1716 the corporation agreed to "a lease to be granted to Mr. Gray of land in the lane leading to Mr. Norris' Mill towards the Gibbet to build a mill [the subsequent Gallows Windmill]." The watermill was then used as a seed mill, and so remained till 1750. The pool is marked on the map of 1785, lying between the three Gallows windmills.

‡ Birkenhead watermill, known in the last century as Bidston Mill, on Wallasey Pool, Cheshire.

15. Townsend Windmill. Action against Burgesses, 1587.

thinketh, whiche milne as this deft thinketh hathe had longe contynuance. And the said Giles Brook saithe that he likewise hathe and occupiethe one other horse milne in the said towne as he verelie thinketh he lawfullie maie.

And the saide defts saie that the said Willm More hathe also wth[n] the said towne one horse milne of his own, and p'tendinge to have some assignement from the said Sir Richarde Molynex K[nt] of one of her Mt[ies] supposed saide winde milnes of Leverpool called Townesend Mylne, hathe p'cured the said Sir Richard to compleyne (as theis defts thinke) in this honorable courte, seeking therby colorably to suppresse and pull down all other horse milnes wth[n] the said town, but beinge (as theis defts thinke) bothe as lawful to be kept uppon and maintained as the horse milne of the said W[m] More and also as necessary and p'fitable for the saide towne, hopinge therby to drawe the p'fitt of the moulture of malte of the whole towne of Leverpoole to his owne horse milne and to gett the whole p'fitt therof unto his owne hand as theis defts thinke.

In respecte wherof the saide defts humbly besecheth this honorable courte to have consideracon of the p'misses and to take suche order both in regarde of the necessitie of the said towne and the extremities the inhabitantes are some times dryven unto for want of water milnes, as also for the better servinge of her Mt[ies] officers and forces at their repaire thither, that the horse milnes nowe in the ocupacon of these defts being verie necessarie and a gret ease to the saide towne maie be kept upp and used; and that the said defts maie have the alowance of the use therof by thorder of this honorable courte as before they have used. [Alternative pleas follow.]

Without—[they deny] that all inhabitants of Lyverpoole by all the tyme wherof the memory of man is not to be contrary have used and of right ought to grind all their corne and malte spent in their sevrall houses in Lyverpoole at her Mt[ies] milnes of Lyverpoole called Easteham Milne and Townsend Milne to the knowledge of theis defts: And without—theis defts or either of them doe not usuallie grinde in their sevrall horse milnes any graine but onlie malt. And without—that by reason of theis defts horse milnes her Mt[ies] said twoe windemilnes in Leverpoole are so hindered as her Mt[s] farmor ys not able to paie the rent reserved to her Mat[ie] for the same, but theis defts saie that if the said defts maie be her Mt[ies] farmors of the said Estham Mylne and Townsend Milne they will put in sufficiente suerties to aunswere her Mt[ies] Rent for the same in the encrease and for the necessitie and behoofe of the towne, and not onelie allow the saide Mr. Crosse his Auncient Winde Milne but the said W[m] More his horse milne and two other horse milnes with[n] the saide towne and as manie more as this honorable Courte shall think requisit. And without—[they deny] that theis defts or either of them have erected or used any horse milne hand milne or querne with[n] her Mat[ies] Manner of West Derby in the said Bill of Complaint named or ells as before ys declared. And without—regarding any other matter or thinge in the said Bill, all such matters and thinges theis defts and either of them will be redie to averr mainteine and

IV. KING'S MILLS, LIVERPOOL.

prove as this honorable courte shall awarde. And praieth to be dismissed out of the same with their costes and chardges in this behalf wrongfullie sustained.

15. Townsend Windmill. Action against Burgesses, 1587.

The defence set up will be noted to be of the usual kind. The rights of her Majesty's "supposed" mills are denied: as they are windmills only, burgesses in times of calm have ground elsewhere at watermills, and ordinarily have ground at horse-mills; the growing necessities of the town and port have exceeded the capabilities of the queen's mills, and it is impolitic to attempt to confine the citizens to them; William More "pretending to have some assignment from Sir Richard Molyneux of her Majesty's supposed windmill of Townsend," besides owning a horse-mill, is endeavouring to put down all other mills for his own profit. The leaders in the revolt do not, on their own showing, appear as purely disinterested parties: John Bird admits being tenant of a horse-mill belonging to Edward Norres; while Giles Brooke has a horse-mill, "as he verily thinketh he lawfully may": and both being indeed ready to cater for the whole grinding of the town "if they may be farmers of the Queen's mills"; for which (since Sir R. Molyneux states these scarce now yield their own rents) they are willing to put in sureties for payment of the queen's rent, and allow as many horse-mills to exist in the town "as this honorable court shall think requisite." However, their elaborate defence formed no reply to the charge of evading the legal soke of the queen's mills: the usual order affirming the astriction to the latter was made, and the illicit establishments suppressed. About a century later Sir Edward Moore, in the *Rental* compiled for his son, refers to this action, though he states it was taken by Edward, not William More (probably Edward having had

Moore Rental: Ch. Soc. Pub. 1847, vol. xii.

the actual working of the milling estate during the life of his father, William) :— IV. KING'S MILLS, LIVERPOOL.

In my grandfather Edward Moore's time there was two or three proud fellows set up mills, but he preferred a bill in the Duchy showing how our windmill [Townsend] is the King's mill, and the tenants within Liverpool ought to grind there, because he paid a great rent. And after two hearings it was decreed that those new erected mills should be pulled down and fined besides; which was accordingly put in execution, and the mills were pulled down. 15. Townsend Windmill. Action against Burgesses, 1587. Rental, 51, 68.

My grandfather, as he was tenant of your King's mill [Townsend], preferred his bill in the Duchy against . . . . . both of whom erected horse mills in the town: after a great suit there was a decree made that as he was the King's farmer within the King's manor there ought to be no private mills, and those who erected them were fined and both the mills pulled down.*

16. William More died in 1602, leaving his milling and other estates to his son Edward (sheriff of Lancashire in 1620 and mayor of Liverpool in 1626).† The lease of Townsend had then eleven and a half years to run, at the end of which time it seems to have been renewed by Molyneux for another term. This Edward Moore, within a few years, was succeeded by his son John (Mayor of Liverpool, 1633), who developed into the Parliamentarian Colonel Moore, Governor of Liverpool during the siege, 1644; when Eastham Mill was destroyed and Townsend Mill, in which the Moores were so closely interested, was preserved. It was during his holding that, in 1636, one of the burgesses was presented at Port Mote for 16. Townsend Windmill. Carriers introduced, 1635. Moore Deeds, Inq. p. m. Port Mote, iii. 153.

* In 1590 John Byrde was mayor, and in March 1591 presided over an Assembly which fined Giles Brooke for purchasing a cargo of grain intended to be purchased by the mayor on behalf of the town as a "town's bargain." The next year Giles Brooke himself was mayor. In 1611, while an alderman, he was presented as finable at the Assembly "for suffering a millstone to lie so long at his door in High Street to the annoyance of his neighbours."

† 1616. Whereas upon complaint made by Edward Moore, esq., that one John Williamson, miller to the said Mr. Moore, had stolen a certain quantity of malt from his master; Upon the reading and perusing of the examinations taken before the mayor and before the said Mr. Edward Moore, it appears that Lucy the wife of Henry Parcoline, Margaret the wife of Robert Whiteside, and Dorothy the wife of John Deane, have received and receipted malt which was stolen by the said miller; It is agreed upon by the said mayor and Assembly here that the said three women shall sit in the stocks at the High Cross upon a market day by Mr. Mayor's appointment, and that they shall put in sureties for their good behaviour, and not be permitted to brew or sell ale hereafter; and Whiteside and his wife shall avoid the town between this and the Annunciation next. Ibid., ii. 808.

16. Townsend Windmill. Carriers introduced, 1635.

"a tussle with the new carrier at Townshende Milne." The old custom of people carrying their own corn and flour to and from the mill was fast dying out both on the continent and in England, and millers were generally giving free carriage. In Liverpool, Crosse, at Middle Mill, seems to have been the first to make the innovation; and Moore, at Townsend, was compelled to follow suit, as stated in the *Rental*:—

Text, III. 134.

Rental, xii. 68.

> [We are now] forced to keep a carrier, the which is a man and a horse more than ordinary. In former times there was never any carrier, and then it was as good a windmill as most in the county; but since Mr. Crosse set his windmill to tenants, they, to get custom, set up the device of a carrier, and we to keep custom were forced to have one too. So that if ever you could buy Mr. Crosse's mill you need keep no carrier, for this reason you must know, they must then either grind with you or be without. . . . You are at the charges of at least £12 per annum upon the maintenance of a man and a horse for carrying.

In 1648–49 the colonel distinguished himself by sitting as one of the judges of Charles I., and signing the death-warrant of that monarch; dying himself the next year, and leaving an estate, which had been little cared for during his life, heavily encumbered with mortgages and debts; and a personal reputation among his fellow-townsmen (who in 1660 promptly became Royalists) that was more than sufficient to utterly condemn beforehand the popularity of his son and heir, Edward.

17. Townsend Windmill. Sir E. Moore, 1650–1678.

17. Sir Edward Moore (created baronet in 1675) entered at the age of sixteen years into a heritage of trouble, that eventually, despite careful economies, parsimonious management, and honest living—despite, too, appeals first to the Parliamentarian Parliament, and then to Charles II.—utterly overwhelmed him and his family. Embroiled alike in debt and in ill favour as the son of his father,* and in disputes with the

Moore Deeds, 61.

Rental, 32.

* "Men who were my enemies gave it out underhand that I was the son of my father, and therefore durst not give their votes for me."

burgesses and the corporation as the holder of Townsend Mill, he passed an embittered, soured, and comparatively short life; during which hard circumstances pressed upon him severely and (as he says on one occasion in connection with his mills) "made me sit down with this great wrong."

17. Townsend Windmill. Sir E. Moore, 1650–1678.

In the hope that his son (fated, however, to die before him) might, on inheriting the estate, gain better results from it than he had done, Moore minutely and laboriously compiled a *Rental* or *Survey*, in which, among many other matters, he referred largely to the administration of the mills upon which he set so great store. It commences with a prayer written in 1667; the year when, in the midst of his trouble, the corporation made a distraint at Townsend Mill. On this unhappy occasion "John Hoole, milner, did upon the 16th day of October imprison both the late bailiffs in Mr. Moore's milne when they were making a distress there"; and Moore had to pay a fine of forty shillings in which his faithful wag of a servant was promptly mulcted by Port Mote. In 1668 the rent of Townsend Mill and the horse-mill being debited by the corporation to John Moore, Esq., shows us that they were now mortgaged; the mortgagee being a London relative of the family.

Moore Rental: Ch. Soc. Pub., 1847, vol. xii.

Port Mote, iii. 854.

Ibid., iii. 871.

18. Under circumstances of so untoward a character Moore hopelessly abandoned the task of attempting to enforce the soke of Townsend in the usual way by actions at law; and, in truth, with his disastrous experience, well might he pray—as in one of his letters to "my onely deare," his wife, June 1670—"God blease alle men from suites." Instead, then, of resorting to suits, he adopted the expedient of securing the custom of at least the tenants on his own private estate; inserting in leases of lands and tenements a clause that the holder should grind at

18. Townsend Windmill. Tenants' Leases.

IV. KING'S MILLS, LIVERPOOL.

18. Townsend Windmill. Tenants' Leases.

one of his mills—*i.e.* at Townsend, or the horse-mill, or Bootle mill. Two or three copies of such leases by Moore remain in the Moore deeds. Taking that granted to Nicholas Rochdell, November 25, 1671, the clause enforcing compulsion and stipulating the penalty for its non-fulfilment is found to be as follows:—

Moore Deeds, 435.

> Provided allwayes and upon condicon, and ye said Nicholas Rochdell for himselfe and his assignes doth covenante p'mise and grante to and with ye said Edward Moore his heires and assignes and ev'y of them by theis p'sents, that when and so often as hee ye said Nicholas Rochdale and his assignes shall have occacon of grinding anie manner of corne, mault or other graine, during ye terme and termes aforesaid, y^t then and soe often as they or any of them shall grinde ye same at some of the milne or milnes of ye said Edward Moore his heires or assignes and not elsewhere: And if hee or they or anie of them shall make default therein, then to forfeite and paye, for ev'y Winchester measure or bushell soe grinded awaye, to ye said Edward Moore his heirs or assigns one shilling of lawfull money of England toties quoties noīe pœnæ. . . . Provided allways that if it shall happen the above menconed yearly rent, or any p'te or p'cell of the forseid [multure] of the said millnes bee behind &c. then it shall be lawful &c. for ye said Edward Moore &c. to distrain upon the said premises.

The framing of leases in this form had become a very general affair throughout the kingdom, and obviously was a commonplace precaution against mills being deserted by tenants of the lands of the millowners. Picton, the popular topographer of Liverpool, however, following the example of Heywood, the first editor of the *Rental*, has fallen foul of it, and chosen to extend to our day the obsolete prejudice and ill-feeling of two centuries ago against millowners; inflicting on the memory of the unfortunate Edward Moore a stigma that only a perfect misapprehension of the facts can palliate. One or two isolated extracts from the *Rental* regarding leases and "penalties" evoke from Picton the absurd remark: "Machiavelli himself could hardly have improved on the hard, astute, cunning selfishness Moore displayed." But let us see. Let us gather together, not one or two only, but all Moore's

Memorials of Liverpool, 1875, i. 113.

observations on the matter, and even make the case against him as complete as possible, that it may be fairly adjudged :—

18. Townsend Windmill. Tenants' Leases.

Rental, 11. 20. 23. 50.

> Mark well the covenant in your leases for grinding at your mills. Let your penalty be as high as you can agree with the tenants. . . . Know every week what tenants you have that grind away without lawful cause, and make them pay for it according to the covenants in their leases. . . . Make your leases according to my new lease in Moore Street. . . . Be careful of the clause to grind at your mills: it is a great thing to your estate, and mind your tenants observe it well. This remember, that those who live upon any part of my lands, ought and are bound by special covenant to grind at some of my mills. . . . Observe well how you dispose of your lands in the field, for there is scarce a man in Liverpool that is not Mr. Crosse's tenant [and bound to grind at his Middle Mill] but upon consideration of leasing to them but two lands would be bound to grind at your mill, by which means you might make your horse-mill of great value. Vide the rule for the horse-mill in every lease you make; for those which have no houses under you but only lands in the field, and live in other men's houses, will under pretence of that not grind at your mill. So that you must advise with counsel for a particular clause or proviso in your leases to bind or force such to grind with you; or else, as I find by experience, they will find some way to avoid you and not grind with you.

These, then, were the instructions Moore thought it necessary to give his son for ensuring custom to the mill. The clause in the lease was not only legal, but was in ordinary operation throughout the kingdom; the "penalty" for absence from the mill being the equivalent of the usual fine and confiscation for such default; or the ordinary payment by the tenant for exemption. Many such leases made by others than Moore are to be found. In 1676, at about the time the *Rental* was written, for example, William, Earl of Derby, leasing land at Wavertree to one Robert Carter, stipulated that the latter "shall always bring all the corn and grain which by him or his assigns and every of them shall be spent yearly in or upon the premises, to be ground upon or at the said earl's mill or mills within the manor of Wavertree and not elsewhere without reasonable cause"; while "if the said R. Carter

Text, III. 236, 250.

IV. KING'S MILLS, LIVERPOOL.

18. Townsend Windmill. Tenants' Leases.

or his assigns, occupiers of the premises or any part thereof, shall fail or refuse to bring all the corn and grain, which by them shall be spent in or upon the said premises, to be ground upon or at the said earl's mill or mills," then the lease shall be void. And, a century after Moore's time, the Hon. C. L. Mordaunt, of Halsall, in the same county, in leasing several closes of land in Down Holland to William Balshaw, of Alcar, stipulated, as usual, that Balshaw and his heirs "shall grind or cause to be ground all his and their corn and grain that shall be yearly used and spent on the said premises for his and their own family's use; and all other the oaten groats and meal that shall be yearly grown upon the said premises, and be by him or them ground into flower or meal, shall be ground at the mill or mills of the said Charles Lewis Mordaunt, his heirs and assigns, in Halsall, and not elsewhere; and in default or neglect of every such grinding as aforesaid shall and will forfeit and pay unto the said C. L. Mordaunt, his heirs and assigns, the sum of Five Shillings of lawful money, to be levied by distress and sale as for rent for arrear, *without suit at law*; if wind and water serve." But to revert to Moore; when his son, Sir Cleave, succeeded to the Liverpool estate, his leases were modelled on the same plan; and with the framing of them was closely associated one of the foremost conveyancing lawyers of his age, Isaac Green, of Liverpool (whose daughter inherited a large fortune in local landed estate and mills, which eventually passed in marriage to the Marquis of Salisbury). In 1698 Sir Cleave, in conjunction with his relative Sir John Moore, of London, mortgagee, granted a lease on land in the new-made Cable Street to Robert Bibby, with the usual stipulation :—

The said R. Bibby, for himself, his executors, administrators, and assigns, doth covenant that he, his executors, administrators, and

IV. KING'S MILLS, LIVERPOOL.

18. Townsend Windmill. Tenants' Leases.

assigns (inhabitants of the said messuage), shall and will, during the term hereby granted, grind or cause to be ground all his malt, corn, or grain which shall be used in or for the respective family or families of the said R. Bibby, his executors, administrators, and assigns (inhabitants of the said messuage for the time being), at some mill or mills of the said Sir John Moore or of his heirs and assigns, being within three miles of the said hereby leased messuage; and in default thereof yielding and paying—and the said R. Bibby doth for himself, his executors, administrators, and assigns, covenant with the said Sir John Moore, his heirs and assigns, that he the said R. Bibby, his executors, administrators, or assigns (inhabitants of the said messuage), shall and will yield and pay to the said Sir John Moore, his heirs and assigns—for every Winchester bushel of corn or grain which the said R. Bibby, his executors, administrators, or assigns (inhabitants of the said messuage for the time being), shall grind or cause to be ground elsewhere than at such the mill or mills of the said Sir John Moore, his heirs or assigns, the sum of two shillings and sixpence, and so proportionately for a greater or lesser quantity; and that within ten days after such default shall be made in grinding at such mill or mills as aforesaid.*

There was, therefore, nothing of "Machiavellian hard, astute, cunning selfishness" in Moore's framing of leases; and Picton's ill-judged censure is singularly out of place. Moore's anxiety that his son should make the most of his legal rights at the mills—the horse-mill, "God bless it! a thing of great concernment to your estate," and Townsend mill, "the king's mill, your ancestors built it"—is very keen, and, viewing his circumstances, somewhat pathetic; yet, save in one instance to be noted, he advises neither injustice nor illegality, nor ever steps out of the well-beaten legal track of centuries of millowners before him.†

The one cardinal sin of Moore as a miller and a burgess—due to the illegal withdrawal of his

* As late as 1814 the Earl of Sefton's legal advisers pursued the same course; stipulating in a lease of land at Sefton to Robert Birch and his heirs that under a penalty they "shall and will grind and shell all their grain at Sefton Mill." This, a watermill beside the churchyard, an ancient establishment near the early seat of the Molyneux family, still remains, and is worked by a descendant of the above Robert Birch.

† This, the first attempted appreciation of the character of Sir Edward Moore, outlined in the paper already mentioned, read before the Lancashire and Cheshire Historic Society in March 1896, has since been followed by Mr. W. F. Irvine in an introduction to his reprint of the *Rental*; and future investigations on the same theme will, it is believed, place the character of Moore in a very different aspect than that in which Heywood and Picton chose to depict it.

Text, ante, p. 125.

Liverpool temp. Charles II. (1899).

IV. KING'S MILLS, LIVERPOOL.

tenants and the political laxity of the times—is revealed in the following extract:—

18. Townsend Windmill. Tenants' Leases. Rental, 50.

Know who are they that grind at your mill that are none of your tenants, in order that if there fall an occasion in your power you may show them a kindness. Know who are they of any fashion that doth not grind at your mill: and (unless they be obliged by some special obligation to them that owns the mill where they grind, as by kindred or the like) I charge you never trust them: neither do them a courtesy if it be in your power. Where you find a great brewer that is none of your tenant, and that doth not grind with you, try if they will be your tenant for one or more lands in the field; and for the same, oblige them to grind with you. Observe the rule above said exactly, and *if ever you be mayor* and a justice of the peace in the country you may very easily make this mill [the horse-mill] worth 20 measures a week, which at a crown a measure is £5 a week, many of your tenants brewing 30 measures a week. . . . My grandfather, his continual allowance in the house [for consumption] was 16 measures a week in malt and 16 measures of bread corn: and he got it all for toll. But it was because he was the only man in these parts in all the great offices, so that all for fear or love grinded with him; and he observed exactly the rate above said. I find at Bootle Mill his usual get was 16, 18, or 20 measures a week. Therefore serve God and follow his example: so you may make your mills worth some hundreds a year, according as corn bears a price.

"If ever you be mayor!" Painful and, in this connection, degrading words were these for Moore to write—an honour once frequently enjoyed by his wealthy ancestors now sought in sordid hope of profit. From the very dawn of mayoral life in Liverpool no family had so often provided occupants of the civic chair as had the Moores; from the middle of the fourteenth to the middle of the seventeenth century their constantly recurring names showing that members of the family had held the honour fully forty times.* But Sir Edward's father

Rental, 87.

Early Mayors of Liverpool: Elton. Trans. Lanc. and Chesh. Hist. Soc., 1902.

* Of this rare distinction Moore did not fail, elsewhere, to remind his son:—"This you may boldly and truly say, the corporation and you have lived together this four hundred and odd years, and in all that time have been in great affection one to another; and not one generation of so many hundred years but your ancestors have been mayors; many of them, in man's memory, two or three times apiece; and one, Thomas de la More, in Ric. ii., was in his life twelve times mayor, as you may see by your Deeds. And this you may further say of truth, which few, if any, in England of your quality can say: that there hath not been a Parliament in 250 years but one of your ancestors hath been [representative] burgess for this town." [According to the deeds, Thomas del More was mayor no fewer than sixteen times.]

(the regicide colonel of 1649), who had been mayor in 1633, proved the last of the long series; neither Sir Edward himself, nor his son—for whose sake he ventured upon crafty counsel that has imperilled his own reputation—nor yet any other descendant of his ever did become mayor.

IV. KING'S MILLS, LIVERPOOL.

18. Townsend Windmill. Tenants' Leases.

19. Townsend Windmill. The Horse-mill, 1361–1667.

19. The horse-mill of the Moores had long been a cherished possession of the family. Originally established by John del More before 1361 to take the place of one of the two watermills, it was situated, not near the site of the latter, but near Moore Hall, at the north end of the town, on the sand-lands, or "white acres," * by the shore; of which the Moore family owned a considerable tract. A deed of 1361, that mentions "a certain road in le quitakrastrete which leads to the mill," identifies its site. Moore Deeds, 194. White-acres Street became known as Milne Street (now Old Hall Street), and is mentioned in various early deeds.† The horse-mill was used for grinding malt for the Moore tenants, and in this guise of a private mill John del More seems to have purchased it from the Duke; as after 1361 it never appears to occur as a possession of the duchy, and is always found to be worked by the Mores. Early in the reign of Henry VIII., William More, finding this malt mill too far from the centre of the town, seems to have established another on his central land in Fenwick

* N.D. (*c.* 1257). Le Witacris. Ibid., 270 (9).
1308. Le q'takirfeld in territorio de Lethirpol. Ibid., 119.
1331. Le Quitaccris. Ibid, 89.
† 1374. Le Milne Street in Lyverpull. Ibid., 233.
1515. Indenture between David Gryffyth, maire, with his choburges, the comynaltie and William More that the latter shall grant a certain right of way for the burgesses, which way lies straight up the Milne Strete and so north into the fields: in exchange for Mill Hill lane, leading out of Milne Strete; which More shall have. Ibid., 341.
1535. A certain way called le Milnestrete, with the lands of the said William More on both sides. Ibid., 354.
1539. Whiteacrestrete or Mylstrete. Ibid., 356.
1588. More strete *alias* teithbarne streete [disproving the modern assertion that Whiteacres Street was Tithebarn Street]. Ibid., 400.

IV. KING'S MILLS, LIVERPOOL.

19. Townsend Windmill. The Horse-mill, 1361–1667.

Moore Deeds, 357.

Alley, and at the inquisition following his death in 1531 he was found to have been possessed of "two horse-mills in Liverpool." Subsequently the mill in Whiteacres Street was abolished and that "in Castle Street" retained; this being the particular horse-mill which the defendants in the cause of 1587 stated William More possessed in Liverpool. In the time of Sir Edward Moore its site may be identified by the *Rental*:—"You may build a house in Phœnix Alley (where Boulton the miller lived) fronting to Fenwick Alley, only remembering to be careful not to stop the lights of the horse-mill. . . . Remember Andell (of Phœnix Alley) doth not with his back buildings stop away any of the lights belonging to the horse-mill." The map of 1765 shows these alleys, and enables the site to be fixed as near the junction of Fenwick Street and Brunswick Street, beside the corner of the Corn Exchange, the approach only to the mill being in Castle Street.* The business done at this malt mill in the centre of the town must have been considerable, and Moore was by no means unappreciative of its value :—

Rental, 50.

The horse-mill, God bless it! a thing of great concernment to your estate. I have got, when the trading to Lockabar was used, 20 measures of toll a week for two years together when malt sold for 5s. a Winchester measure. But now 9 or 10 measures a week, and against fairs or holidays 20 measures a week.

This remember, have a great eye how custom rises or falls at this mill. Know every week what tenant you have that grinds away without lawful cause, and make them pay for it according to the covenants in their leases. . . . Mark well the covenant for grinding at your mills: let your penalty be as high as you can agree with

* Near the mill on the east side of Castle Street stood the once well-known hostelry "The Millstone." In 1738 the *Lancashire Journal* advertises the loss of a horse at "The Millstone at Liverpool"; and in 1744 one Ralph Peters charges the corporation, according to their books, with "Expenses at the Millstone taking affidavits 8s. 6d.," in connection with one of their lawsuits. It was then "an inn of the first class," and one of the few at which post-chaises were kept. In 1766 one of the two stage-coaches "which go constantly to London in three days in the winter and two in the summer" started from "The Millstone in Castle Street"; the other running from "The Talbot" in Water Street. Herdman says the inn was kept by one Elizabeth Harrison "as late as 1766"; but ten years after this date the Liverpool newspapers advertised a sale there of the effects of George Cockshott; the last host who presided over the fortunes of "The Millstone."

them. It would be a good way to invite, one day in the Christmas, all that are customers to your mills, and be known to your tenants both in the town and country. I am confident it would get both this mill and the other great custom.

IV. KING'S MILLS, LIVERPOOL.

Remember there can be no more mills in the town than what is already.

20. Encroachments on what soke remained to him gave Moore some concern :—

20. Townsend Windmill. Encroachments, 1667.

Of late Captain Fazakerley hath erected a mill in the Castle which he pretends is not in the liberty of the town. Query, if you may pull it down, since it is none of the king's mill nor pays him no rent? Or query, how there might be an order made in the mayor's court to see who doth suit at the king's mill [Townsend], and so to procure an order that the tenants belonging to the king may not go from the king's mill out of the liberty of the town.

Rental, 70.

Captain Fazakerley was governor of the royal castle of Liverpool, in which, as already mentioned, there had been a horse-mill and a bakery for the use of the garrison from very early times till the year 1663. At this date "a true and perfect account of what part of the Castle of Liverpool is to be demolished" included the item—"a house between two breadths of the inner wall that, pulled down, must likewise destroy the horse-mill." But it had not been destroyed, and, according to Moore, was grinding for the burgesses; claiming to be on private royal estate, and therefore outside the liberties of the town. In any case its operation for burgesses was certainly illegal; for the king before this date had sold his manorial rights in Liverpool, and his mill could not be justified in encroaching on the soke he had sold. The "king's tenants" Moore mentions were practically his own, he being the king's miller, and they had no right to frequent the Castle Mill. Except for the disturbed state of affairs and Moore's troubled condition, he would have had no difficulty whatever in suppressing the operations of the Castle Mill by an action at law.

Lord Lieutenant's Minutes, 1662–76.

Text, I. 124; III. 228.

He was also perturbed as to a diversion of the ancient stream from Moss Lake on the heath, that

IV. KING'S MILLS, LIVERPOOL.

20. Townsend Windmill. Encroachments, 1667.

in former days had driven Eastham Mill. It still drove a very small watermill that stood half a mile farther up than Eastham, (Norris' Mill, beside the Gallows Windmill); but practically it had become useless as a mill-stream till Lord Molyneux, owner of Moss Lake Fields, by diverting it in another direction, contrived to make it drive two new watermills in Toxteth Park. Moore, complaining of the political disadvantages that had prevented him protesting against this, urged that the stream should be turned back again from Toxteth Park, and so cause the closing of the mills there:—

Rental, 69.

Remember one other thing of great concernment. Within the memory of man the Lord Mullinex hath erected two watermills in Toxteth Park [outside the boundaries of Liverpool]. . . . Whereupon considering it was just at the king's restoration, and that all these red-letter men were so high, and the chancellor of the Duchy was near akin to him [Molyneux], and likewise the lord treasurer of England was his brother-in-law [&c.]; all which considered made me sit down with this great wrong. If the water be taken off the Moss Lake it will be better for your windmill [Townsend] by £10 per annum, for that it will make the Park Mills want water, their greatest supply being from that lake.

21. Townsend Windmill. Profits of Multure.

21. Of Townsend, the core of the entire milling estate—"the windmill you hold in fee-farme of the king"—he enjoined:—

Rental, 68.

The Windmill. This you hold in fee-farme of the king at the rent of 30s. a year, which, considering that your ancestors built it, and are forced to keep a carrier, the which is a man and a horse charges more than ordinary; which together with the daily repairs and casualties is as much near as the mill is worth or more, had you not so many tenants of your own which are great bread bakers and are bound to grind with you at some of your mills.

There must be no mills but this in Liverpool.

Remember there can never any more mills be erected within the manor of Liverpool but what are now in being; for this reason, yours is the king's mill, and you pay him a great rent and are at the charges of at least £12 per annum upon the maintenance of a man and a horse for carrying, and at least £5 per annum for repairs, and £9 per annum board wages for a miller: which in all [with 30s. rent] amounts to £27 10s. per annum that must go in ready money out of purse yearly: besides, if a casualty by a stress of wind come—which often falls out—the mill may be damnified 10, 20, 30, 40 or more of pounds.

21. Townsend Windmill. Profits of Multure.

Remember the greatest improvement I can advise you. Put £100 to his in your windmill that now William Gardiner hath. Two mills together would bring you £20 in the £100 for the building another mill. Cause another windmill to be erected, for you will need but one carrier and one horse to them both, and a miller and a boy of a cheap wage about 16 years old. In doing so, and making the water run the right way [so as to shut down the Toxteth Park mills], you may easily make your two windmills worth £80 per annum or more. For my grandfather hath formerly set this mill for £27 per annum: but now the carrying and the Park mills takes off so much that, all charges borne and the mill kept in as good repair as I gave it to them, I have now during the lease but £10 yearly and two bushels of wheat at Christmas for it. But if there were two mills, considering the great store of tenants you have in the town and that one carrier would serve them both, truly it is not to be thought what advantage they would bring you, especially observing the rules to your customers which I have set down in the directions for your horse-mill; therefore vide horse-mill in Castle Street. Old rent of this mill £10 and two measures of wheat yearly.

22. Townsend Windmill. Sale, 1724.

22. The whole of Moore's quaint reflections, all his sage advice for building permanently up a valuable milling trade in the town, were but wasted labour; for his eldest son died before he was able to utilise them, and a younger son, who succeeded, failed in the heavy task. With the one exception, provoked no doubt by the laxity and difficulty of the times, they are creditable alike to his sagacity and his sense of justice at a very critical period of soke history. Now that this period has happily closed, if we complacently look around us to-day and smile at his solemn injunction—"Remember there can never be any more mills in Liverpool than are now in being"—let us congratulate ourselves that, though there are more mills, there is no more soke, no more compulsory astriction of tenants, no more suppression of free competition, and no more filching away of the rights of owners of mills.

But the troubled race of Sir Edward Moore was well-nigh run. In 1677 the corporation obtained from Lord Molyneux a one-thousand-years' lease of the fee-

IV. KING'S MILLS, LIVERPOOL.

22. Townsend Windmill. Sale, 1724.

farme of the town; * Moore's superior lords now being the commonalty that had ever been active in hostility to his mills, and that, as he finally must surely have foreseen, would ere long render all his advice for their preservation needless. In 1678, at the age of but forty-four years, the last and most notable of the king's millers of Liverpool, following his ancestors and his ancient predecessor in the craft, William the son of Adam, was buried within the fane of St. Nicholas' by "the Sea Bancke called the White Acres"; leaving his younger son, Sir Cleave Moore, heir to the mortgaged estates and mills.

Moore Deeds, 390

For some few years longer Sir Cleave continued the struggle, but, except a trifling item in 1672 regarding the working of the mill on Sunday,† we hear nothing of Townsend. In 1689 the distressed owner witnessed the first overt act of the corporation for the abolition of the ancient soke in the town:—

Port Mote, iv. 260.

"Upon the second petition of William Hands he is granted a lease of three lives for Thompson's croft: his former fine of £4 remitted, and twelve pence yearly rent to continue, provided he lays out £100 upon the premises upon a Windmill, and continues it in good repair, and so leaves it at the expiration of his lease." This was rapidly followed by similar grants, and in a few years the ridge of high land extending southward from Townsend Mill and Eastham became dotted with windmills possessing no soke, and catering in open competition for the grinding of all and sundry who might resort to them. With the further development of the town, other new mills in all directions covered the slope of the heathside; till, by 1768,

App. to this Chapter.

* The reversion of this lease the corporation purchased in 1777, and became absolute owners of the old royal estate.

Port Mote, iv. 12.

† 1672. June 3. We present Henry Bannister for grinding corn on the Sabbath day at Middle Mill; Richard Mercer for the like offence; Edmund Buckley for the like offence at Townes End Mill; Robert Woolfall for the like offence: each being fined 3s. 4d.

no fewer than twenty-seven windmills (besides one watermill), then marked in the map of Liverpool, must have imparted to the aspect of the place from the riverside much of the semblance of a Dutch landscape.

IV. KING'S MILLS LIVERPOOL.

22. Townsend Windmill. Sale, 1724.

In 1724 Sir Cleave perforce gave up his impossible task. Sir John Moore, the mortgagee, foreclosed; and the whole of the Moore estates in Liverpool, Kirkdale, and Bootle, including the Moore Hall in Liverpool and Bank Hall in Kirkdale,

Bank Hall, the Seat of the Moores.

were sold. Sir Cleave left Lancashire for the South of England, and died in 1730; and though three succeeding baronets maintained the name of the family, none of them retrieved its estates; the heritage of the eventual heiress, the daughter of the last baronet, Sir William, being comparatively little more than mere charters, rolls, and deeds bespeaking antiquity and former wealth—the now oft-quoted "Moore Deeds." Yet within half a century of the sale by Sir Cleave "these lands, houses, and tenements increased wonderfully in value, and bring in

Hist. Liverpool: Enfield & Perry

IV. KING'S MILLS, LIVERPOOL.

22. Townsend Windmill. Sale, 1724.

more per annum at the present time [1773] than they cost [at the last sale]; and it is well known that sales have been made of parts of them at sums equal to five times over the first fee-simple." Bank Hall, with its moat and towered gateway; its open-timbered roofed hall bedecked with busts and shields; its courtyard wall, "whereon was a grand arrangement of all the armorial acquisitions of the family on shields elevated like a battlement, and the date 1282"; as well as most of the Kirkdale and Bootle estates in its neighbourhood, were purchased by the Earl of Derby.*

Lord Derby in 1724 further purchased "All that messuage and tenement in Oldhall Street in Liverpool commonly called the Old Hall, also that Windmill situated and being at the lower end of Dale Street in Liverpool commonly called Townsend Mill";† which latter, with a score free-trade mills around it, no longer retained any pretence to its ancient exclusive rights, and humbly ground on lease for hire. Its aspect in 1772 is taken from an engraving in Enfield and Perry's *Liverpool*; the site being on the west side of the Wellington Column at the top of the brow, near the Art Gallery.

Frontispiece to this Chapter.

23. Townsend Windmill. Destruction, 1780.

Port Mote, xii. 49, 137, 185.

23. In 1777 the corporation purchased from Lord Molyneux the reversion of their thousand years' lease of the town; and subsequently one of their earliest acts was to instruct the town clerk "to write to the stewards at Knowsley about the purchase of Lord Derby's interest in the Townsend Mill; and also Mr. Rigby, the present lessee, concerning his interest

* In 1773 "this venerable dwelling-house hath lately been demolished, and will probably be soon forgotten." Its site is covered by Bank Hall railway station and sidings, in the street of the same name.

† Conveyance consulted by consent of the Earl of Derby at Knowsley. Here is also preserved a large oil painting of the time of Sir Cleave Moore, showing an extensive panorama of the Moore estates, from the timber post-mill of Spellow to the mills of Bootle Marsh on the Shore.

also; in order to pull the same down." Two years later it was "Ordered that the leasehold interest in the mill commonly called Townsend Mill and the houses and premises there, belonging to Alderman Rigby, be purchased by this corporation at the price of £300, on this condition only, that the corporation shall be able to purchase at such price as they shall think right to give the reversion of all the said premises from the Earl of Derby, in order that the nuisance may be removed and the premises laid to the highways there for the benefit of the town." Finally, in 1780 it was agreed that "this corporation do purchase the reversion of the site of a mill, cottages and waste land adjoining, at the east end of Dale Street, at the top of Shaw's Brow, from Lord Derby for £300." Shortly after, "the nuisance," rich in associations with venerable antiquity, was destroyed; the £600 paid by the corporation in respect of its site and trade being the only payment ever made or expense ever incurred by the corporation in connection with the extinction of the milling soke of Liverpool.*

IV. KING'S MILLS, LIVERPOOL.

23. Townsend Windmill. Destruction, 1780.

* Apparently the term "Nuisance" was applied to the mill because of the danger apprehended from its sails revolving close to the highway. Edward I., at Winchester, narrowly escaped death on one occasion, by his horse being frightened by the sails of a windmill—strepitum ventrincolæ. A modern instance in point is amusing:—"An indictment was tried at Huntingdon which excited no small degree of pleasantry in the county. A clergyman indicted a miller for working his mill so near the common highway as to endanger life. The clergyman is a man of considerable property and consequence in the county. He was obliged daily to pass by this road on horseback, and had been several times thrown by his horse taking fright at the sails of the mill. The mill was an old one which had formerly stood on a common, but the latter had been enclosed by the Commissioners under the Inclosure Act, who had directed the new highway to be so placed that it passed close under the fly of the mill. Mr. Justice Grose, addressing the jury, said the mill as it now stood was unquestionably a nuisance, and the miller must be found guilty. It was, however, no fault of his. The Commissioners who directed the road to be set out were most to blame, and he regretted they had not been made parties to the indictment. Neither was the prosecutor to blame in preferring the indictment: he could go by no other road, and his life as well as those of his fellow-subjects was endangered. Under such circumstances he felt wholly at a loss how to act. The miller ought not to be punished; yet the safety of the prosecutor and the public must be consulted. He thought the best way of deciding would be to direct *the prosecutor to pay the miller £40*; and the miller to abate the nuisance, with leave to erect his mill on some convenient spot adjoining. This was accordingly made the decision of the Court"—apparently with the ready assent, if not at the suggestion, of the good-natured and well-to-do prosecutor.

Leyland MSS. i. 181.

Annual Register, July 18, 1807 (402).

IV. KING'S MILLS, LIVERPOOL.

24. Rural Mills.

Text, ante, p. 162.

24. In the action at law, 1587, Sir R. Molyneux pleaded that he was lessee of "five windmills within her Majesty's Manor of West Derby, commonly Eastham Milne, Townsend Milne, Ackers Milne, Wartre Milne, and Derby Milne." The first two were literally Liverpool mills; the other three were situated four or five miles from the town; but the whole five belonged to one common owner—the king, Duke of Lancaster. Two of these three rural mills, those of Accers and Wavertree, were annexed to the town in 1475. In that year the king granted to Henry Crosse, gentleman, "the town and lordship of Lytherpole with the appurtenances" (which included Eastham and Townsend mills) at the old rent of £14 per annum; "and also one windmill called Akersmyle and another windmill called Wartre Mylne,"* at a further rental of 20s. each. At the close of that year the king's receiver included in his receipts:—

Close Rolls, Edw. IV. 79.

> The town and lordship of Lytherpole with appurtenances: also one windmill called Ackersmyle and another windmill called Wartre Mylne in the county aforesaid, parcel of the duchy: For the town and lordship £14, and for the mill called Akersmyle 20s., and for the mill called Wartre Myle 20s.

In this form the two continued to be interpolated in the leases of Liverpool till the sale by Charles I. The amount of their rentals in 1475, as compared with those of Eastham, 28s. 6d., and Townsend, 13s. 4d., in 1450, proves that they were fair-sized structures, doing a good trade in a rich wheat-growing district; and their addition to the fee-farme of Liverpool must have been considered a considerable augmentation of the milling resources of the growing town. Their rentals of 20s. each remained the same in all the leases; though in 1528, when Henry Ackers and Alice Griffiths secured the lease, the rent of the town and appurtenances (including Eastham and Townsend) was raised from £14 to

Duchy Grants in Fee.

* "Wartre" is still the popular local colloquialism for "Wavertree."

24. Rural Mills.

£14 6s. 8d. Ackers never worked the mills; but after leasing the town and town mills to the corporation, let Accers and Wavertree for all his term at 46s. 8d.; of which 40s. was payable to himself, and 6s. 8d. to the royal bailiff of West Derby. The rental of the town, as sold by Charles I. in 1629, was £14 6s. 8d.; and therefore the sale did not include these two mills, which were reserved and sold separately.

25. Accers Mill.

25. Though not added to the town leases till 1475 Accers Mill is mentioned as early as 1342. In that year, at an inquisition respecting the forest of West Derby, the verderers reported that two oaks had been cut down in the park of Croxteth for the repair of the Mill del Accers, "but for the most part they were old and dry"; also that one sapling oak had been cut down in the outer wood to make a mill-shaft for Accers; while in the time of William of Penrith, seneschal of the forest, three oaks had been cut down for the repair of Accers Mill.* Unfortunately the identity of the mill in this connection has been obscured by the trivial clerical error of the transcriber, Okill, misreading its designation as "Atters," owing to the well-known similarity of the letters *c* and *t* in ancient manuscript; Baines and other local writers subsequently adopting the inaccuracy.

Okill MSS., xv. 84 (1).

Hist. Liverpool, 1852, 150.

At this early period of its history Accers Mill was leased with the neighbouring rural mill of West Derby. In 1349 the bailiff of that manor responds for two windmills, and the same appear there again in 1430; while in 1446 the compotus of Henry Norris, royal steward for West Derby Manor, definitely includes receipt of rental and payment of expenses at Accers:—

* Quod duo querc̄ prostrat' fuerant in parco de Croxstath hoc anno p' reparacione molend del Accres sed erant p' majore p'te sicce & vet'es. Et quod unū sapplynge querc̄ prostrat fuit hoc anno in forinseco bosco quedam milnshaft faciend' apud le Accersmilne, et tres querc' p' reparacōe molend' del Accers.

Inq. Forest, 16-17 Edw. III. 8.

25. Accers Mill.

Close Rolls, 23-24 Hy. VI.

Customary Repairs: Carriage of one oak from West Derby Wood to the windmill called Accursmylne, two leagues, 8d.; purchase, for the same, of a square [iron socket for insertion of the four arms] and fittings for a shaft at the said mill, upon which to place the lattice-work for the sails, 3s. 4d.; wages of one man half a day, repairing with pales and posts certain breakages in the fencing, 1s. 1d.

Receipts: 24s. for the farme of the windmill called Acoursmylne, as demised to William Mercere by Sir Richard Molyneux, seneschal there; to be held by the said William from the Feast of St. Michael, in the eighteenth year of the present king, for the term of nine years then ensuing; this year completing the sixth. The said William to sustain and repair the said mill in all things during the aforesaid term at his own proper costs and charges; [the lord] to find him structural timbers, but not the cutting or carriage of the same, and millstones, with their carriage, to the said mill. The said William has had in the first year of the said term payment of 20s. for repair of the said mill, undertaken outside the agreement for the said term, and excluded from this account on the testimony of the said Richard Molyneux upon the account; the previous term of John Wright, former farmer of the said mill, not being then formally completed by him, nor his sureties, Henry Wilkinson and others, brought forward. And the said William is to deliver up the said mill in sufficient repair at the end of the said term.*

After 1475, when the lease of the mill was added to that of Liverpool, the rent was fixed, as it remained, at 20s. In 1546, when Sir William Molyneux and his son Richard obtained the lease of the town, they also obtained possession of the rural mills; and in 1557 the compotus of Sir Richard Molyneux, then bailiff of West Derby, thus accounts for the rent of Accers:—

Sheriff's Accounts, 4-5 P. & M.

Receipts: 20s. for the farm of the windmill called Acresmill, as demised by indenture under the seal of the Duchy of Lancaster, given November 25, in the thirty-seventh year of King Henry VIII.,

---

* In carr. unius quercus de bosco ib'm usq' molin ventriticum voc Accursmylne p' ij leucas viij^{d}: et in empcoē ejusdem ead'm causā squarr' apt' p' uno shafte inde habend ad id'm molin in qua pecten velā imponuntur iij^{s} iiij^{d}: et in stipendū unius hoēs p' dies di cum palis post et railez rep'antis certas rupturas palicii xiij^{d}.

De xxiiij^{s} de ferma moli ventritici voc Acoursmylne sic dimis' Willmō Mercere p' Ric'm Molyneux milite sen^{l} ib'm. Habend Willmō a festo Sci Michis anno xviij regis nunc usq' ad finem tmū ix annoz extunc p'x sequenciū et plenarie complend' hoc anno vj^{to}. Et id'm Willm's sustent' et rep'abit d'cm molin in om'ibz durante t'mino p'dicto sumpt' suis p'prius et expn̄: quod [*blank*] ei invenit grossum meremiū absq' p'stracoē et cariagio, et molares cum cariagio eoz'dm usq' d'cm molin. Et p'dcus Willm's h'abit in primo anno t'mi p'dci de xx^{s} p' rep'acoē d'ci moli exceptis p'exceptis durante t'mio p'dco excludendo p' testion̄ p'dci Rici Molyneux sup' comp'm non obstante t'mio Ioh'is Wright nup' firmar' d'ci moli nondum complet eo qd id'm Ioh'es non p'fecerit suam p'positū Henricus Wilkynson et alioz. Et id'm Willm's dimittet d'cm molin sufficient rep'atum in fine t'mi sui p'dci.

to Sir William Molyneux and Richard Molineux, his son: to have and to hold to them and their assigns from the Feast of St. Michael the Archbishop last past before the signing of the said indenture, for a term of twenty-one years then next ensuing and to be fully completed; rendering therefor 20s. per annum at Easter and St. Michael's Day in equal portions: the said farmers and their assigns to sustain and repair the said mill and leave it in good condition. With a clause for re-entry and without any reservation. This year is the eleventh of their term.

IV. KING'S MILLS, LIVERPOOL.

25. Accers Mill.

The mill was sub-let afterwards at a fair profit; as, for example, in 1589—"Accers: for the mill there called Accers Milne, from John Cole 33s. 4d., out of which is paid to the Crown 20s."; and in 1616, when it yielded Molyneux 45s. per annum. Subsequent to the sale by Charles I. in 1629 it appears in the Molyneux rentals till 1651, when it seems to have been again sold; and it was then worked by various tenants, in what precarious trade they could best procure, till the early part of the nineteenth century.

Much doubt has been entertained as to the site of Accers, its mill, and its hall, the seat of the ancient family of the same name.* Baines considered it was at Manchester, and Holt and Gregson quote Salford as its locality; notwithstanding that the family were of Liverpool, and that the mill (which, owing to the misreading of its name already noted, they had not identified) was included in the Liverpool leases. But the hall stood in Finch Lane, West Derby, on the site of the present farmhouse known as "Ackers Hall." It was conveyed by Henry Accers in 1562 to Sir R. Molyneux, and was the dower-house of the Lady Molyneux whom William More married. Here

Lanc. & Ches. Hist. Soc., 1880.

Hist. Lanc., 1866, iv. 72.

H. & G. MSS., iii. 138.

* The family name occurs in early undated deeds. In 1320 John del Accers witnessed one of the Moore Deeds. In 1333 the Exchequer Subsidy Rolls for the district contain the name of Richard del Accers. In 1342, at the inquisition on West Derby forest, one seasoned oak was stated to have been granted to John del Accers. In 1529 Henry Akers (or Accres), husband of Alice, widow of David Griffith, mayor of Liverpool, we have seen to be joint lessee of town and mills. In 1576 Captain George Accers, son of Robert Accers, of Accers Hall, co. Lanc., was granted coat armour: in his will he is described as "of Liverpool, gent."; and in 1590 his arms appeared in one of the painted windows of St. Nicholas' Church, Liverpool, where he was buried.

IV. KING'S MILLS, LIVERPOOL.

25. Accers Mill.

Liverpool Port Mote, ii. 116.

the latter was living at the death of his father in 1575; when, at the inquisition—"We find William More, as he the day of this enquiry is called William More of The Accers near West Derbie, to be the right heir of the late John More of the Bank Hall." On the tithe map of the Croxteth estate in 1837 the mound upon which the mill stood is marked in a field numbered 2195 and named "Millfield." This was on lands held by the yeoman family of the Boltons, whose house, an interesting structure dating from the early part of the fifteenth century, stood hard by, and was only taken down as recently as about six years ago.* Accers Mill was probably leased by the Boltons from Molyneux till 1589; as in that year the latter is found, on the one hand, leasing Accers Mill to John Cole, and, on the other, leasing Eastham Mill to John Bolton; the Eastham lease being held by the Bolton family till 1606. The Bolton tenement at Accers was owned in 1837 by one John Tarbuck; and here was born very early in the century Margaret Mason, a late aged resident in the neighbourhood, who remembered as a girl the ruins of the mill existing on the mound in "Millfield." In 1881 the mound was levelled by the tenant of the farm, who discovered among the débris broken millstones and timber.

26. Wavertree Mill.

26. The other of the rural mills in Crosse's lease of 1475, was that of Wavertree. Beyond its constant recurrence in the town leases till the sale of 1629 little record seems to remain of it.† Clearly it was a profit-

* Mr. R. D. Radcliffe, F.S.A., purchased the excellent oak timbering of Bolton's tenement, and erected therefrom the handsome memorial Lych Gate which now enriches Knotty Ash Church, Liverpool. A curious recessed oak canopy, which extended across a daïs at one end of the hall, Mr. Radcliffe also purchased and presented to the Free Public Museum, Liverpool.

† Possibly some information might be gleaned from the early Court Rolls and other records of the manor of West Derby preserved by the lord of that manor, the Marquis of Salisbury, at Hatfield. The late Marquis recently kindly gave permission for a search to be made, but unfortunately no opportunity for doing so occurred.

WAVERTREE MILL, LIVERPOOL.

IV. KING'S MILLS, LIVERPOOL.

26. Wavertree Mill.

able concern. Sir William Molyneux leased it in 1546, and it was held by Sir Richard in 1557, and in 1587 it was one of the five entered in the pleadings. Though its Crown rental was but 20s. per annum, Molyneux's tenant in 1598, James Forster, paid him 46s. 8d., the highest rental of any of the rural mills; while in 1648 Earl Derby's tenant, William Keikwith, paid the still higher rent of £5 per annum. It had passed into the possession of the Earl of Derby in 1629; and in 1676, as already related, he bound his Wavertree tenants by the terms of their leases to grind there. Later the estates and mill were acquired by Isaac Green, of Liverpool, solicitor, (the legal adviser of Sir Cleave Moore, of Townsend,) by whose daughter they passed in marriage to Bamber Gascoyne, M.P. for Liverpool. In his time was passed the Act of 1768 for dividing and enclosing the common lands of Wavertree; the portion allotted to him including that upon which the mill stood. By clause 23 of this Act, "If any person, &c., shall erect or build any house or building or shall plant any tree or trees within the distance of two hundred yards from a certain windmill situate on the common hereby intended to be enclosed; or shall suffer any tree or trees, planted without the distance aforesaid, to grow to such a height as to prevent the going of the said windmill, the same shall be and is hereby declared a nuisance, and shall and may be removed or prevented by the same Bamber Gascoyne or the owner or occupier of the said windmill." Various general Acts of this period referring to windmills will be found to contain similar provisions for the preservation of this then valuable form of mill.

Text, ante, p. 171.

8 Geo. III., sec. 23.

Legislation such as this in former days would have relieved many a windmill owner from the necessity of moving his mill from one site to another; and in the present case the Act may perhaps be

Text, II. 272, 321.

IV. KING'S MILLS, LIVERPOOL.

26. Wavertree Mill.

credited with incidentally preserving Wavertree Mill on its ancient site. For, despite the prohibition, within a few years' time tall trees in the thickly wooded Heathfield Park, which fringed the very edge of the little mill-croft, began to cut off the prevailing south and south-west winds ; causing Foster, the miller, to complain to his landlord, Bamber Gascoyne. The latter, knowing he had legal redress available, instead of contemplating moving the mill complained in his turn to the offender, Colonel Bourne, of Heathfield Hall ; whereupon the irate colonel—resolved neither to have his trees lopped, nor to be grumbled at by the humble miller on the other side of his park fence—promptly leased the mill himself and silenced the unfortunate Foster. Ere long the colonel was laid in the neighbouring churchyard ; and Gascoyne also died, leaving the estates and mill to his daughter ; by whom they were conveyed in marriage to William, Marquis of Salisbury, in the possession of whose descendant, the present marquis, they remain. Though the mill had lost its soke rights, it was worked till near the close of the nineteenth century ; the last tenant of any moment being Charles Taylor, who held it from 1859 to 1875 ; its capacity during this period being, with two pairs of stones, six bushels of wheat or eight bushels of coarse grain per hour. It was permanently closed in 1889.

This, "the windmill called Watre Mylne" of 1475, enjoys unique distinction in being the only one of the many ancient mills in and around Liverpool which still exists. In much the same form as it was then, and on its original site near Wavertree Church, this mill, which by the modern extension of Liverpool is now comprised within the boundaries of the town with which it had for centuries been so closely associated, is one of the three edifices in Liverpool which

WAVERTREE MILL, LIVERPOOL, 1896.

still remain on their ancient sites, unchanged from ancient uses; the others being the mother church at Walton, and the chapel of St. Nicholas by the riverside. Thus fortuitously does the vast city of modern times seem to preserve that association of ancient church and mill which innumerable little Saxon hamlets of Domesday maintained centuries before the earliest days of Liverpool.

26. Wavertree Mill.

Text, II. 101.

The abandoned mill stood for several years, more or less a ruin, exposed to the ravages of the weather and the devastations of vagrants of the countryside; still, fragile structure as it seems, it weathered with little additional damage a heavy gale in February 1895 that wrecked many an apparently stronger edifice. Since 1896, when the accompanying view was taken, and a plea entered for its preservation in *The King's Mills of Ancient Liverpool*, the late Marquis of Salisbury happily caused the fabric to be repaired and took ample measures to secure its safety for many years to come. The interest investing the relic is not that alone of the antiquary; but of the people at large, whose corn is soke free and whose bread is abundant and cheap—thanks to those vast changes in political economy and industrial progress that have revolutionised the quondam state of affairs typified by this obsolete mill.

27. West Derby Mill.

27. We come now to the fifth of the mills which in 1587 Molyneux pleaded he held under the Crown—namely, West Derby windmill. Unlike Accers and Wavertree, it was never included in the town lease, but was held separately by Molyneux and others; though to all intents and purposes it was a Liverpool mill, doing a great part of its trade with the town and port. Like the others, it was an extremely ancient establishment. It was included in the West Derby possessions of Edmund Plantagenet in 1297—"two

IV. KING'S MILLS, LIVERPOOL.

27. West Derby Mill.

mills, one a windmill and the other a watermill; worth by the year five marks"—precisely the same milling estate, and of identically the same value, as that he held in Liverpool. Also at West Derby he held "an old castle" and a meadow of twenty acres; while "the tenants of the town render to the lord for use of a certain way through the middle of the meadow twelve pence yearly"—items proving useful in the identification of the site of the mill.

Chancery Rec., Inq. p. m.

Of the watermill nothing more is known, as it early disappeared; but memory of it long survived in the place-names "Millpool" and "Milldam Hey." At the inquisition on the death of Earl Thomas in 1326 it was stated he held a field of ten acres called the Millpool; and in 1796, in the admittance of the Rev. Glover Moore, rector of Halsall, as copyhold tenant of certain land near West Derby Church, occurs the Milldam Hey, or field. In 1826 this field, among others, was purchased by Lord Sefton, and added to Croxteth Park. A small brook exists in the locality; and across the adjacent Meadow Lane * is a culvert, through which it is believed the old mill-stream still runs.

The windmill of 1297 occurs also in the inquisition of 1326; the earl having then possessed "a windmill, let from old time at 100s. a year; and a watermill, let from old time at 40s. a year"—higher rentals even than that of Eastham Mill, Liverpool, as late as 1440. In 1348 the bailiff of West Derby accounts for the rents of the twenty-acre meadow, two windmills, and a horse-mill. The two windmills would undoubtedly be those of West Derby and its nearest neighbour, Accers Mill; and they appear, again in conjunction with the horse-mill, in a similar compotus of 1430.

Harl. MSS., 6461. 28.

In 1446 the accounts of Norris, bailiff, afford some glimpses of West Derby windmill and horse-mill:—

* Formerly called Castle Lane.

IV.
KING'S MILLS, LIVERPOOL.

27. West Derby Mill.

Close Rolls, 23-24 Hy. VI.

Manor of West Derby. Compotus of Henry Norris, bailiff there for Edward Crosse, deputy seneschal, 23-24 Henry VI.

Receipts. 43s. 4d., farme of twenty acres of meadow, called Earl's Meadow; 12d. for a certain way called Milne Gate juxta the chapel of West Derby; also 40s. for the farme of one windmill and one horse-mill, as demised to Richard Penketh for a term of ten years which have expired in the year last past.

Repairs made according to custom. Perforating and grooving one Paris millstone bought for the mill called the windmill of this manor, in place of another much worn in the twenty-second year of the present king [1443]; also bedding the said stone: in all, 5s. Iron purchased for making the point of the iron fusil [mill-pick] of the said mill, 9d. 600 nails called board-nails, bought by computation to fasten the woodwork of the said mill, at 4d. per hundred, 2s. Also idleness of the said mill during the time of this account while under repair in the middle of the season and totally unable to grind, according to the certificate of the seneschal and others affixed to this account; six weeks at the rental of 30s. per annum 3s. 6d. [paid to the miller]. Total for the windmill, 11s. 3d.

A certain agreement for a lump sum made by computation with a carpenter that he should restore and replace with oak timber the arms of the wheel of the horse-mill appertaining to this manor, which are broken, and also repair other defects in the woodwork of the same, 4s. Fashioning one old millstone for the said mill, grooving and bedding it, as agreed, for the sum of 2s. Iron purchased to make the point of the iron fusil of the said mill divers times in the term of this compotus, 12d. One saw purchased to recompense the said mill for another one broken, 4d. Total for the horse-mill, 7s. 4d.*

After 1475 West Derby Mill was no longer conjoined with Accers. In 1476 a writ of Edward III. for the

* Man' de Westderby. Compūs Henrici Norreys p'pos ib'm p' Edw' Crosse deput' sen 23-24 H. 6.

De xliij^{s} iiij^{d} de fir. xx acr. p'ti voc Erlesmedowe. De xij^{d} de quad'm via voc milngate juxta Capellam de Derby. Et de xl^{s} de ferma unius molend' ventritica et unius moli equini sic demiss' Rico Penketh ad t'mū decem annoz que p'terunt in anno ultimo elapso.

Cust' repac'. In p'foratoē et verberatoē unius paris petrar molarium p' molio vocat ventritico hujus man'ii loco alt'ius paris fere p'usitat in anno xxij rege nunc empt' una cum cubacoē eand'm petrar in grosso v^{s}. Et in ferro empto p' fabricacoē finis fusilli ferrei d'ci moli ix^{d}. Et in DC clavoz voc bordnaile p' computē empt' p' opibz ligneis d'ci moli figend' C^{m} ad iiij^{d}, ij^{s}. Et in stacoē d'ci moli p' vj septimā infra tempus compī dum erat in rep'ando nullo modo medio tempe molentis p' certificacm̄ sen^{li} et alioz sup' hunc comp'm jux^{a} ferme de xxx^{s} p' annū iij^{s} vj^{d}. Moli ventr' xj^{s} iij^{d}.

Et in quad'm convencoē p' computē in grosso f'ca cum uno carpentario ut ipse cum m'emio quer' restauret et imponet brachia et alias pecias m'emii in rotam moli equini hujus man'ii p'prius fere defract iiij^{s}. Et in aptatoē unius vet'is petre molaris p' eod'm molio cum verb'atoē & cubatoē ejusd'm p' convenc' in grosso ij^{s}. Et in ferro empto p' fabricatoē finis fusilli ferrei ejusd'm moli div's vicibz p' temp' compī xij^{d}. Et in una serura empta p' hostio d'ci moli loco alt'ius defracte iiij^{d}. Moli equini vij^{s} iiij^{d}.

repair of Liverpool Castle also directs certain repairs at West Derby Mill:—

27. West Derby Mill.

Item: Ye wyndmyll at Derby next to ye horsmyll is well repaired in all things except bordynge and shyngling, which is ordeyned to bee doon. The Hows of the said horsmyll is soe feble yt it can not stande, wherefore it is orddeyned to bee taken down and a new Hows to bee made, whereunto is assigned the Tymbre of ye old stable at Lyvpole without ye castell, and such other old Tymbre wtin the castell as may bee spared, and the residue to be had at Symondeswode, and carried.

Duchy Orders, Edw. IV. 100.

In 1546, when Sir William Molyneux obtained leases of Accers and Wavertree Mills, he obtained also the lease of the windmill and horse-mill at West Derby, together with twenty acres of field in the Erlesmedowe, and pasture of the herbage of the castle fosse, for twenty-one years, paying therefor 79s. sterling: viz. for the mills, 26s. 8d.; for the twenty acres, 43s. 4d.; and the fosse, 9s.; he to keep the mills in repair, and to have permission to move the said horse-mill to any other desired site in the manor. The rent of the two mills, which in 1444 had been 40s., will be noted to have decreased to 26s. 8d. The Earl's Meadow was the twenty-acre meadow of Edmund Plantagenet, Earl of Lancaster, in 1297. It stretched down the slope from the windmill towards the village; and through the middle of it ran "the certain road called Milne Gate" (because it led to the mill gate), for using which, in 1297 and again in 1444, the tenants are seen paying twelvepence per annum. This road is still called Mill Lane. In 1557 Sir Richard Molyneux, as royal bailiff, accounts for the rental of the windmill and horse-mill. In 1568 William the Miller paid Molyneux £4 for the year's rent of the mill; but later it and the Earl's Meadow were let together at £10 per annum. Queen Elizabeth renewed the lease in 1586, and the mills were held by the Molyneux family till, in 1609, they

Duchy Leases, II. xxx. 177.

27. West Derby Mill.

proved to be first of the Liverpool mills alienated by the Crown; on May 29 of that year James I. including "a windmill and horse-mill in West Derby with appurtenances, value 31s. 8d. per annum," in a batch of about one hundred mills in different parts of the country, which formed one instalment of his wholesale sales of mills to the large speculators in these properties, Edward Ferrers, mercer, and Francis Philips, gent., both of London. In due course a Molyneux repurchased the mills, and, apparently abolishing the horse-mill, enlarged the windmill; this subsequently appearing in the rentals of the estate down to the close of the eighteenth century. It remained leased out at the rental of 31s. 8d. till 1767, being then raised to 40s., at which it stood till 1796, when it was abolished.

Text, III. 19.

Croxteth Register of Leases, 1607–1796.

The site of the structure may still be identified, being shown on Perry's map of the Liverpool district of 1768, and on a map of the Croxteth estates in the following year. It stood near the main road leading from Liverpool to the village, overlooking from its eminence Mill Lane, the ancient way called Milne Gate juxta the chapel of West Derby.

28. *Appendix:* Their Immediate Successors.

28. Referring, in conclusion, to milling affairs within the borough towards the close of the seventeenth century, the ancient soke is found to have been entirely broken by the action of the corporation granting their building lease in 1689 to William Hands for the erection of a windmill, and quickly following this with various similar leases for the same purpose. Private owners of land in the borough soon did the same; and within a century mills developed with so marvellous a celerity that corn milling in even that age assumed the position of one of the foremost industries of the town. Windmills were most suitable for the locality; and it was amid a group of the sturdy, squat towers of these that the old peg-mill of Townsend ended its days. Within these primitive evolutions of the art of mill-building, an excellent trade, for that period, was conducted by individual owners, whose energy founded the nucleus of the vast milling business now characteristic of Liverpool. In their day, without doubt, many interesting little histories clustered round these now long-vanished mills, many a story of enterprise and success; for on the whole the period was one of vast prosperity; but scarce a

IV. KING'S MILLS, LIVERPOOL.

28. *Appendix*: Their Immediate Successors.

tradition of these or yet a relic of the mills remains. In the year 1768 Enfield's map of Liverpool shows no fewer than twenty-seven of these mills.* Subjoined are a few notes of the principal establishments that thus took up and splendidly developed the trade of the ancient king's mills of Liverpool; and among them are some personal recollections kindly imparted by a former member of the local craft, Mr. George Lunt. These may perhaps be considered an appropriate memento of an order of things that in its turn has passed entirely away.

Mills on the North Shore.

Several mills under various ownerships stood on the north shore of the Mersey.

"Townsend Mill" (modern). By the courtesy of the Lancashire and Yorkshire Railway Company, I find from their title-deeds of this site that in 1742 Peter Furnival leased a certain plot of land adjoining Mill Lane North to one James Wilcock. The land passed to Wilcock's son-in-law, John Adamson; who on November 19, 1792, renewed the lease of "all that piece of land situate or being on the strand or shore of the river Mersey, on part whereof the said John Adamson hath erected or is erecting a wind corn mill." This was the modern Townsend Mill; which of course does not appear in the maps of 1768 and 1785, and is first seen in Howarth's map of 1803. Its tall tower, seven stories high, prominently erected near high-water mark, formed a conspicuous object on the long line of sand hills and gorseland at the entrance to the river. Adamson seems to have left Townsend in a few years and started in Bridgewater Street the first steam corn mill of Liverpool, which was burned down in 1819. About 1834, after the destruction of Bootle Windmill, Jeremiah Shaw occupied Townsend, where he met with his death in a somewhat singular manner. The most terrific storm of the century at Liverpool occurred on Sunday and Monday,

Text, II. 295.

* Townsend Mill, almost in the centre of the map, at the junction of Folly Lane (Islington) and London Road, may readily be identified. Adjoining it on the north are two windmills at the top of Clayton Street. Eastward are two mills in Folly Lane, and close by the three Gallows Mills in London Road.

Returning to Townsend, four mills, slightly to the south, are ranged along Lime Street on the site of the L. & N.W.R. Company's station. One in the rear is "The White Mill" of Copperas Hill (also on the site of the railway station), associated with a mysterious tragedy. To the south-east are two mills in Mount Pleasant and Brownlow Hill, adjoining the former being the inn where was born William Roscoe, the Florentine historian. Near the Observatory is another. Farther south still is a group of three, standing on the site of the future Cathedral. Below, towards the river, may be discerned one in Park Lane; and, almost on the shore, a tide-mill, beside a pool fed by the Toxteth Park stream.

Once more returning to Townsend, the two Middle Mills are seen due west: three Bevington Bush (or Scotland Road) Mills to the north, the tall tower of one of which, still standing, is used as a warehouse for the modern roller-mill adjoining: one westward, near the corner of Pinfold Lane (Vauxhall Road) and Tithebarne Street: and, at the extreme north near the river, a single one; this being the modern "Wishing-gate Mill," near which, twenty years later, was built the new "Townsend Mill."

Enfield and Perry's "Liverpool."

MAP SHOWING LIVERPOOL WINDMILLS IN 1768.

IV. KING'S MILLS, LIVERPOOL.

28. *Appendix*: Their Immediate Successors.

January 6–7, 1839, causing an unprecedented series of wrecks in the river and off the port, and the loss of over a hundred lives. Mr. Lunt, referring to this storm, says: "In the great gale of '39, Jerry's mill, on the edge of the river, could not be held in. It was usually reckoned indeed to be the best blown mill at the north end of Liverpool. During the storm Jerry was on duty all night, and every sack of wheat had been shot on to prevent her taking fire. But it was all to little purpose, and she tore away in spite of every effort to check her. There was a place in all windmills to which a brake could be applied on emergency, and at last Jerry put this brake on, and to steady it sat on the end of it. This was a long beam of stout wood. He managed to keep the mill in check during the night in this way; but the strain and shaking he received were very severe and he never shook off their effects; and not long afterwards they proved fatal. I very well remember Jerry, as he was always called, from my earliest days. My father acted as his executor. The late Thomas Bolland was his foreman and married his daughter, carrying on the mill for some years after Jerry's death." The last tenant was the late Frederick Dresser (of Edmund Street Rice Mills), who worked it for grinding rice husks from 1872 to 1880, the wind power being abolished in favour of steam. It was burnt down, December 12, 1880: this being not altogether an unfeared catastrophe for some years—as Mr. Dresser stated, "it was a dangerous risk, and I paid three guineas per cent. premium." The lower part of the tower, now deeply embedded among warehouses and docks, in Regent Street, opposite the Bramley Moore Dock, is used as a cement warehouse; adjoining it are left two or three of the antique little cottages that in former times were occupied by the millers.

"The Wishing-gate Mill." This is shown in Perry's map, 1768. It stood in a sand-land adjoining the shore, about five hundred yards nearer to Liverpool than Townsend Mill, and overlooked the popular Wishing-gate, whence it derived its name. Hither a century ago, and later, were wont to repair friends and relatives of seamen on board outward-bound vessels passing seaward close inshore. In those days of small ships, long voyages, and months of silence at home, many a sad scene of farewell was witnessed at this old mill on the sands. Its site, measured from the old church, by reference to various maps, seems to have been at the north-west corner of the present Stanley Dock.

"The Coffee-house Mill." Still about four hundred yards nearer Liverpool was a mill standing in a shore field belonging to Lord Derby, and shown in the maps of 1785 and 1803. Reference to the town books for February 17, 1809, shows that the mill had been purchased by the corporation in 1802 from Thomas Plumbe, and that in 1809 it was burnt down. On May 12 in that year Edward Lyon, lessee under the late Thomas Plumbe, was ordered to surrender the lease and make good the damage. In 1810 Robert Greenham was the occupier, and in 1811 he was called upon "to pay £150 damages due to the fire." It obtained its designation from an adjoining coffee-house. Having been restored, it was worked till

1821, when it was again burnt down. As late as 1830 this spot, not more than three-quarters of a mile from the Town Hall, was the favourite and fashionable bathing resort, where, says Herdman, "the beauty of the shore was not to be surpassed at any modern watering place." The coffee-house, with its tea-gardens and bowling-green, was kept at this date by a Dutchman, one Vandrie, whose name is perpetuated in that of a neighbouring street. The open shore, thronged, in Herdman's pretty drawing of the scene, with bathing-machines and parties of holiday folk, is now covered by the Clarence Dock.

"Townside Mill," nearer Liverpool still than any of the others, in 1785 overlooked the "intended new gaol," subsequently "The French Prison." On June 29, 1789, the Liverpool papers reported a fatality here: "On Saturday, the 27th inst., a violent storm of rain and hailstone passed over the town, accompanied with the most awful and tremendous thunder and lightning ever remembered. Great damage was caused in the town, all in the space of a quarter of an hour. At the *Townside* Mill belonging to Mr. Humphrey Green, situate near the new jail, great damage has been done. The lightning shivered the upper shaft and proceeded down the sack chain to the upper working floor, where it forced out the whole of a leaden casement. It then got to the lower working floor, where it overturned several large iron weights, forced out another leaden window, and made its passage through that floor into some rooms below which are occupied by the miller's family as a dwelling. In one of these, used as a common sitting-room, a young man was thrown against the grate. In the next room were two children, one of whom received a stroke in the eye, and they were both nearly suffocated by the sulphurous vapour. In another room was a girl 12 years old, daughter of Mr. Green, who was instantly struck dead. The window of this room and the next were materially damaged and a hole made through the door. The son of Mr. Green, in attempting to get the children out, was twice repulsed by the sulphurous vapour before he could succeed.—It now appears that it was not Mr. Green, the owner of the mill, who lived there and whose daughter was killed, but Robert Winstanley, a poor miller, who has suffered repeated misfortunes."

Lime Street Mills.

Upon the long ridge of high land extending along the heath from ancient Eastham southwards were erected several mills, the sites of which are now covered by Lime Street Station and its approaches. Five in line, on the site of Duncan or Hotham Street, overlooking the site of St. George's Hall, appear in Perry's map of 1768.

"Tyrer's Mill," one of the oldest, and the only one then in Liverpool built in the antique peg style, was doubtless that erected in 1689, as already stated. Tyrer's Mill verged on the north side of Copperas Hill, where subsequently was erected the Roman Catholic Cathedral. In 1744 occurs mention in the town's-book of "Benjamin Tyrer's house, mill, and field, next to and west of the White Mill

IV. KING'S MILLS, LIVERPOOL.

28. *Appendix:* Their Immediate Successors.

field"; and in 1781 "Anne Tyrer, widow, was granted leave to alter the lives in the lease of Windmill cottages on Copperas Hill." In the interim had occurred a tragedy which remained undiscovered for fifty years. The White Mill and Hadwin's Mill close adjoined Tyrer's; and the tenant of the White Mill, who had formed a partnership with the tenant of Hadwin's Mill, was murdered and buried under this latter structure by, it is alleged, his partner. Evidences of the crime were discovered about 1820, when the mill was taken down for the erection of "The Quaker's School," a building which has given way to tramway stables. The story, as related in the Underhill MSS. in the possession of the corporation, is as follows: "Duncan Street.—Here was the site of an ancient [modern] mill belonging to the late Joseph Hadwin [of the Society of Friends]. A free school was built upon the site, and when the land was required for the purpose, the workmen were directed to demolish the mill in two days; but, if I am rightly informed, they accomplished the work of destruction in two hours. In the centre of the mill, and deep in the rock upon which it stood, a grave was found with a human skeleton lying within it. Mr. T.'s [Tyrer's] aunt, upon hearing of the circumstance, hastened to the spot, and instantly formed an idea of the cause of this singular interment. Many years ago, perhaps fifty or sixty, this mill and the White Mill, a short distance above it, were occupied by two partners, one of whom resided at this mill [Hadwin's]. He was an unprincipled character. His partner, who bore the name of a decent, upright man, left his home one morning to go to the town, but his wife never heard of him again. Nor could any one explain the cause of his sudden disappearance until this discovery was made, when his wife and child and his partner had long been dead."

The owners of the field in 1785 were Waterworth & Co., and in 1790 Pennington and Pait were partners there. In the same mill, in September, 1790, took place the following example of endurance under disaster: "On Friday George Drover, who takes care of a mill near Limekiln Lane [Lime Street], belonging to Messrs. Pennington and Pait, went into the lower apartment to grease the wheels without taking the precaution to stop the mill. His arm was caught in the wheel and torn off near the shoulder. He then went up six or seven steps that lead to the bank outside to stop the mill, which, having done with his other hand, he went again below and sat there some time before he called to some rope-makers in the neighbourhood. He was taken to the infirmary, and was so stout [hearted] that he would have walked. The stump of his arm was amputated, and he has every appearance of doing well." The old peg-mill of Tyrer lasted till February, 1793, when, "in the storm of Saturday se'nnight, the wooden mill on Copperas Hill, that has weathered so many gales, was blown down and totally demolished. Unfortunately, at the time the miller was coming down the ladder, and was buried under its ruins. He was got from it in a short time, and, what is wonderful, without any broken bones; but he has received a very violent contusion on the eye, and it is feared he will lose it."

28. *Appendix:* Their Immediate Successors.

GALLOWS MILLS.

Three mills bearing this designation stood in London Road, near the east corner of Stafford Street.

Reference has already been made to the first, built in 1715, and the execution of the Jacobites there in 1716. At this date the ancient Moss Lake stream was still running beside this mill on its way to Downe Street; and Norris' watermill (of 1587) was still driven by it. The original Gallows Mill and its site were purchased by the corporation for public improvements in 1788, at a cost of £700 (Joseph Gerard being the miller at the time), and was advertised for sale by auction the year following: "The materials of the Gallows Mill on the Prescot Road: to be taken down and removed at the expense of the purchaser," the selling price being £71. Beside it, at the corner of Stafford Street, stood a tavern, which remained an object of some curiosity on the coach road till as late as about 1865.

Text, ante, p. 134.

The second of the three was built in 1719 by Alderman Thomas Tyrer, who had already petitioned the corporation for leave to build a mill in Tithebarne Street. The London Road lease was granted for the erection of the mill by Tyrer for three lives and twenty-one years at 2s. 6d. per annum. By 1757 it had passed to Samuel Jones, and 1774 to William Farrington.

The third, though also built by Tyrer (in 1749), became known as Dobson's Mill, having been purchased (in 1755) by John Dobson, a miller and merchant, of whose bankruptcy and subsequent proceedings, extending over many years, an interesting account might be written.* In 1800 a petition of William Rowe to alter the lives in the lease "cannot be granted until the mill be taken down in order to widen the road"; but in 1812 Edward Blackstock, gentleman (half on his own behalf and half as executor for Edward Newsham), is granted a similar request "subject to taking down in a limited time the said mill, as well as the public-house called 'The Barleymow,' for the widening of the said road." This inn, in an advertisement for its sale in May, 1800, is stated to have frontages on Prescot Road, Pembroke Place, and towards Shaw's Brow; and stood therefore along the angle of the roads in Monument Place. About 1780 this lonely spot on the high road was infested by footpads and highwaymen, the capture of three of them by "the patrol" being triumphantly recorded in 1783.

* "1778, May 8. To be sold by auction by order of the assignees of Mr. John Dobson, at the Golden Lyon, Dale Street, all that parcel of ground and the windmill, two houses, and other buildings thereon erected, and the small garden situate on the north side of the highway or road leading from Liverpool to Low Hill: bounded on the south side by a stone delph and the said highway, road, or lane: containing by estimation 2 roods 12 perches. All the above premises are held by lease under the Corporation of Liverpool for three lives and twenty-one years, under a yearly ground rent of 10s., and were lately let to Mr. Isaac Smith at the yearly rent of £80. They are now in good repair, and very well adapted for a miller on account of their contiguity to the turnpike road."

IV. KING'S MILLS, LIVERPOOL.

28. *Appendix:* Their Immediate Successors.

MILL LANE MILLS.

Opening out from the site of Townsend Mill, at the top of Shaw's Brow, and alongside the Art Gallery, is Mill Lane, which a century ago led to an open croft containing two windmills; this again leading into Clayton Street, down a flight of stone steps placed wholly across the street, some part of which still remains. These mills are shown in Enfield's map of 1768. The one nearer to Shaw's Brow in 1784 was "raised to a very great height at considerable expense," and was the tallest in Liverpool. It had been owned by Thomas Taylor, corn-factor, who lived in the house adjoining, at the corner of Shaw's Brow, and whose death had occurred in 1780.* Noah Smith was the miller here till his death in 1806.

Of the other mill, known as Carson's, there seems little record. In 1803 was offered for sale "the beneficial interest for an unexpired term of thirteen years in a wind corn mill, drying kiln, and garden at Mill Place, near Shaw's Brow, together with three small dwelling-houses and stable adjacent, subject to a clear rent of £65 per annum; also the dwelling-house on the east side of Rodney Street, lately occupied by Mr. John Carson." A view of this mill, in the rear of Christ Church, is published in Troughton's *History of Liverpool*, 1810. The three small dwelling-houses attached to the mill, trim and neat yet apparently, still remain.

The usual fate of windmills befell both these structures; Carson's being burnt down August 10, 1824, and Taylor's April 23, 1852. It is the capless tower of the latter which in Herdman's drawing is seen rising to the rear of the houses on the brow, on the site of the Art Gallery. In 1813–15 these mills were employed in grinding flour and peas for Wellington's army during the Waterloo campaign.

One of the reminiscences of Mr. George Lunt's father in connection with Sunday grinding at Taylor's Mill about 1826 has already been related. The miller there at that time was Richard Rawsthorne, who resided in the house beside the mill, as Taylor had done. "Gerard Street, near the mill," says Mr. Lunt, "was where my father first set up in the baking business. It was at that time a rural well-to-do and somewhat aristocratic neighbourhood. I have heard him speak of hearing the cuckoo when seated at his bedroom window in Gerard Street during a spell of illness. His business increased so rapidly that he was shortly the employer of some half-dozen hands. Those were good old days! Country wheat was

Text, II. 317.

* "1784, Jan. 29. To be sold by auction, all that substantial and well-built corn windmill; together with five dwelling-houses, two stables, a croft, and garden, let (exclusive of the mill) at the yearly rent of £25. These premises are situate on Shaw's Brow, in Liverpool; were lately the estate and in the possession of Mr. Thomas Taylor, corn-factor, deceased; are now in the occupation of Messrs. William and Robert Taylor, millers; and are held by lease under the Corporation of Liverpool for two lives now in being and a reversionary term of twenty-one years, under a small yearly reserved ground rent of 5s. The mill has been lately raised to a very great height at considerable expense, and the purchaser may enter upon immediate possession. Apply to Suddell & Blackstock, attornies, who want £3,000 about on eligible land security."

28. *Appendix:* Their Immediate Successors.

bought at 70 lb. the bushel as now, but there was generally 2 lb. or 3 lb. over to a sack; while with the last load would usually be sent from the farm 2s. or 3s. for the miller, and not unlikely a good fat hare in a sack for 'th' mester.'" *

MOUNT PLEASANT MILLS.

On the broad slope of Mount Pleasant (or Martindale's Hill) stood two windmills. One, situated at the rear of the "Bowling-green" Inn, was approached from Brownlow Hill, and was thus frequently styled Brownlow Hill Mill. It was built by John Martindale, who, in 1737, was granted a lease of "his house at Brownlow Hill and the mill and house adjoining"; the lives in the lease being altered in 1751, Martindale agreeing to pay 2s. 6d. per annum for the mill site. In 1753 William Roscoe was granted a lease of the inn and bowling-green "upon the north side of the road leading to Martindale's mill"; and here, on March 8 of the same year, was born his son, Roscoe, historian of the Medici.

The other of the two was erected higher up the hill, near the corner of Hope Street and the site of the Medical School. It was built in 1717, when Samuel Gilbody and his partners were admitted by the corporation to a plot of land, acquired from John Dowse, there to build a mill and make certain improvements, on payment of a guinea fine and sixpence per annum rent. In 1731 Gilbody, together with Ralph and Hannah Seacombe, petitioned the corporation that "the said mill had lately been blown down, and they had erected two other mills upon other parts of the corporation estate,

---

* "My father was with the miller at Sefton, Mr. Rothwell, as teamster before coming to Liverpool; his duties there being chiefly delivering flour to dealers in town, for whom Mr. Rothwell ground, I believe, at a rate of 6d. per old Liverpool bushel of 70 lb. Oatmeal was also made at a charge in the same way, though I fancy the batching work done for farmers would be paid for in toll generally. This would be about 1820 or 1822. My father's energy and "go" were such that he quickly changed the character of his employer's business, by making actual cash sales of flour and oatmeal in Liverpool. So much so, that the little Sefton mills had quite as much as they could do without hire work, though it would be incumbent on the miller to do what the tenants required in that way. The 6d. rate paid very well, as no risk was involved and power cost nothing. The grinder would probably get a guinea a week or more, and would likely board in the house. (He would get 10s. extra if at a windmill, as he was liable to have to get up at any hour of the night if a wind sprang up, and work on whilst it lasted, without thinking of overtime: nothing was known of overtime in those days.) Sefton Mills, being on the edge of a long stretch of a dead level, were frequently waterlogged by the backing up at flood time, and on that account at busy times the stream stood in much the same position to them as the wind to a windmill, and the miller must work at the ebb, whether by night or day. I have heard my father tell more than once of catching some half-dozen large baskets of grayling in Sefton Mill-race—I wonder how many could be caught there now? He found the country slow for him, and consequently made an engagement with a much-respected flour-dealer—but not baker—of that time, Mr. Peter Melling, say about 1823 or 1824. Mr. Melling was contemporary with the late Mr. Golding, the grandfather of the present Goldings; as also of Mr. Blanchard, and the late Mr. Henry Jump, father of the late Ralph and James Jump; and some others whose names have long passed away."—*Mr. G. Lunt, jun.*

IV. KING'S MILLS, LIVERPOOL.

28. *Appendix:* Their Immediate Successors.

which they were not bound [by the terms of their tenure] to do, although they had agreed to keep their original mill in repair"; wherefore they desired to be exempt from rebuilding the said timber corn mill. On condition that they agreed to maintain the other two mills, they were exempted accordingly.

In 1771 it was "ordered that the mayor and bailiffs let the mill called Brownlow Hill Mill and two small houses belonging thereto, falling into the corporation's hands, for the best terms they can get, but not to let it for any term longer than from year to year. Joseph Brooks to have the preference offered him, and to be let clear of all leys, taxes, and repairs." It was already determined to take the mill down, hence the limitation to an annual tenancy only; but it was not till 1779 that it was agreed, "in order to remove the nuisance of Brownlow Hill Mill, Ralph Astley, the occupant, be at liberty at his own expense to take down and remove and dispose of the materials of the said mill." Astley apparently saw no profit in the speculation, and in 1792 the corporation announced—"To be sold by auction the materials of a windmill situate upon the south side of Mount Pleasant, behind the Workhouse, to be pulled down by the purchaser."

Cathedral Mount Mills.

Three windmills stood on the plateau alongside the huge stone quarry, known then as Zion Mount, later as St. James' Mount, and now as Cathedral Mount, whereon is to be erected Liverpool Cathedral overlooking the cemetery in the old delph. The public pleasure-walk or promenade on the mount, planted with an avenue of trees and affording a charming prospect of the river, the meadows of Wirral and the Welsh hills in the distance, had at its south end a windmill used for making linseed cake and oil; and it was to avoid the second of the mills on the mount from being converted to a similar undesirable purpose that in 1777 the corporation decided "that the mayor and committee purchase at a fair price the windmill and premises on Quarry Hill from Mr. Drinkwater, to prevent its being purchased by others who might convert it to some use that might be detrimental to the public walk." The earliest allusion to these mills seems to occur in 1723, when Edward Appleton "proposes to lay out a great sum of money in building upon and improving a piece of land which, together with the mill and buildings on Quarry Hill, had been formerly ordered [to be leased] to him and John Livesey." In 1766 it was Thomas Appleton's mill, and in 1782 is mention of "Appleton's mill and bowling-green at the top of Duke Street." In 1805 the newspaper notes the death, at his residence, Quarry Hill, of "Mr. Edward Haighton, a truly Honest Miller"; who, alas! had only recently been a bankrupt.

Bevington Bush Mills.

In 1768 three windmills stood here in close proximity to the rural Scotland Road, a fourth being added later. On December 27, 1790, it was reported that "in a high gale on Thursday the top

IV. KING'S MILLS, LIVERPOOL.

28. *Appendix*: Their Immediate Successors.

of a mill at Bevington Bush, with its machinery, was carried off and thrown into an adjoining field." In 1798 were "to be sold a piece of ground near Bevington Bush, with the mill thereon erected, called Paul's or Appleton's Mill; also another plot adjoining, with the mill thereon, called Lund's Mill: this latter subject to a lease granted to Anthony Lund." In 1801 was "to be sold all that wind corn mill on the west side of the lane between Kirkdale and Liverpool, now in the possession of Mr. James Willacey; works four pair of stones, is well built, well situated, roomy, powerful, and in full trade: if not sold will be let to the best bidder." Of these mills Mr. Lunt kindly gives a few practical notes:—

"There were four windmills here. That on the west side of Limekiln Lane was worked by a well-known miller (a great friend of the Rev. John Shepherd), Anthony Myers. He was very successful. So far as I know the family are extinct in Liverpool. Later on these premises were bought by the Huntington family, who are not so old as millers as they are as bakers in Liverpool.

"The mill on the east side, where Simpson's Mill now stands, was worked by Edward Weston, John Simpson, and George Lunt, under the style of Edward Weston & Co. Each of the partners had a retail shop or shops, and they also ground for hire. Edward Weston took to betting, and as a consequence met with misfortune, and the mill was then carried on by John Simpson and George Lunt. I find by an old rent-book that steam was added in 1838, the steam mill being built against the windmill tower. There were two egg-ended boilers and a beam engine. The mill was burnt down in 1850, and the estate was then bought by Simpson and Lunt from the original owner, J. Whalley, of Maghull, near Liverpool. At our engine on one occasion they could not get her to pump, and of course the boilers would steam dry. Our water came from a tank which formed a roof for the slack-hole. Our engineers were sent for, and the pump was pulled to pieces; but nothing was wrong with it. They examined the tank, and found there was plenty of water there. As a last resort, after a day or two's stoppage, they began to examine the piping from the tank, and then found that a coat button had got wedged in the angle. The charge for grinding was never less than 6d. per old Liverpool bushel of 70 lb.; at least we never charged less. There were some who offered to grind at 5d.; but generally one dose completely cured the baker who was venturesome enough to try it, for the stuff had to be eaten. This did for sizing purposes, but for bread, 'Never.' I think Mr. B. Molyneaux (who was the last to grind for hire on the old system) charged 7d. The charge for grinding always included delivery of the products, as the farmer always delivered the wheat into the mill; and it was country wheat which was so ground for the bakers. The miller generally covered and charged insurance up to a moderate amount, but if much excess in stock was held the owner of the wheat had to put a supplementary policy on—as, for example, the policy in Norwich Union Office for £350 negotiated by my father in 1834 for wheat in Jeremiah Shaw's mill at Kirkdale. At balancing time (if balancing time ever came for a miller in those days, for it used to be

IV. KING'S MILLS, LIVERPOOL.

28. *Appendix*: Their Immediate Successors.

an axiom that you never knew what a miller was worth until he died) it was the custom for the miller to furnish what was called a finishing ticket, and you had always a hard job to get it—sometimes indeed, you never got it. It was made up on the basis of 21 packs (240 lb.) fine and seconds, 1 pack thirds, 2 packs fourths, 4 packs 1 sc. 10 lb. bran, 1 pack 0 sc. 10 lb. loss, to the 100 $\frac{0}{70}$ bushel wheat; the difference in quality was made up between the fine and seconds. You might have, say, 20 packs fine and 1 pack seconds, or if you wanted very fine flour, say 18 packs fine and 3 packs seconds. We used to enter our grindings up something in this way:—

Received 25 sacks containing 100 $\frac{0}{70}$ [equal to 29 packs 2 score] white wheat for Mr. George Lunt, ground as below:—

	Packs	sc.	lb.
Superfine	18	0	0
Seconds	3	0	0
Thirds	1	0	0
Fourths	2	0	0
Bran	4	1	10
Loss in screens and grinding ...	1	0	10
	29	2	0

Loss allowed $2\frac{1}{2}$ lb. per bushel of 70 lb.

"My father eventually sold his interest on account of declining health to his partner, J. Simpson, who thus became sole proprietor. About this time silk dressing came into vogue, and Mr. Simpson made considerable alterations, putting in new boilers and engines, with silk-dressing reels; Hind and Lund of Preston doing the work. After Mr. Simpson's death the premises became the property of his son Joseph, who has also passed away.

"A little farther to the north-east stood, and still stands, the mill so long kept by that dear old worthy, Thomas Leicester. How we love the memory of the old man! He always had, I believe, both shop and mill; but as he did not commence until 1827, I don't think he was the first to combine the trades. Both Blanchard and Jump were older than he as millers, and my father a year or two older as a baker, but not quite so old as a miller. Blanchard was amongst the first to adopt baking along with the milling; certainly I have heard him 'blessed' for the introduction of cutting and weighing bread. I have often heard my father say he could do better in the way of profit in the bread trade when the corporation fixed the price week by week than ever he could do since. They generally allowed a margin of 15s. or 16s. per Liverpool peck of 240 lb.; but your bread had to be stamped '2 lb. I. W.,' and if it fell short of weight it was seized. But there was no cutting up of loaves in the shop. It is to Blanchard that the trade is indebted for the cutting up of bread." The tall tower of this mill, rising from the midst of extensive modern roller-mill buildings, is now the only one existing in Liverpool.

"The next of the Bush Mills was worked, and I believe owned, by a member of the Rawsthorne family (James, of Shaw's Brow)."

28. *Appendix:* Their Immediate Successors.

Crabtree Lane Mill.

This mill, which stood on the site of the coal sidings in Falkner Street (Crabtree Lane) about half a century ago was the most perfect steam mill in Liverpool; and its owner, Mr. Stephen White, attained a considerable competence in its management. Mr. Lunt writes: "Mr. White was a very enterprising and keen man, and about the time of the outbreak of the Crimean War, 1854, Fairbairn, who had been concerned in bringing out the then new dress in millstones which caused them to cut shearwise, was employed by the War Department to construct two floating mills for the use of our Army and Navy. These did their work automatically, and were considered the wonder of those days in milling. Mr. White at once saw his opportunity and his man, and I think gave Fairbairn *carte blanche*, or thereabouts, to refit his mill; with the result that steam was installed instead of wind, and from the Moss Lake stream close by a pool was made in which the condenser was cooled. The mill being brought to the very front for the manufacture of baker's flour, gave White's trade an impetus it never lost so long as he was able to give his business personal supervision. His 'straws' held sway as a baker's flour for many years, and for long was unrivalled."

Bootle Mills.

Bootle Watermill on the shore, a little beyond the Liverpool boundary, survived the age of the Moores, and was finally utilised as a paper mill; the later windmill, erected beside it, being used for corn. Reference has been made in an earlier Volume to a pioneer but unsuccessful attempt to work the old water-mill by steam. Neither of these mills endured long after the opening of the last century.* Text, III. 293.

* "To be sold by auction at Bootle Coffee House, near Liverpool, the windmill called Bootle Mill, situated upon the banks of the River Mersey, about three miles from Liverpool, and 300 yards from the canal: also the scite of the Watermill there, house, garden, &c., and land extending to 18 acres. The premises are held by lease under the Earl of Derby for a term of 49 years from December 25, 1792, at a reserved annual rental of £120. The fall of water at the scite of the mill is 37 ft or thereabouts, which is supplied from Bootle Springs, with three cubic feet of water every second for 12 hours in 24, even in the driest season. The weight of this supply is 187 lbs., which will overcome a resistance equal to 1,288 lbs., and will work 1,500 spindles [if used as a cotton mill]. The windmill works three pair of stones, and is in good repair. There are watercourses and millponds annexed, and troughs lately made for conveying the water from Bootle Springs to the watermill; and the proprietors of the canal have agreed to permit the owners of these premises the use of an overflow from the said canal 20 yds. long, and ⅛ of an inch lower than any other from the said canal. The water of the springs is very pure, and of a proper quality for making fine paper, and the situation of the land is a very agreeable one for accommodation for sea bathing, and is already much resorted to by bathers. Fuller information from the Executors of Mr. John Singleton." Liverpool Advertiser, Aug., 1793.

"To be sold all those mills called Bootle Mills, including both the watermill and the windmill, land, &c. Water-wheel 33 ft. in diameter: the watermill lately used as a paper mill. The whole on lease from Lord Derby for a term of which 39½ years are unexpired." Ibid., Sept., 1801.

28. *Appendix:* Their Immediate Successors.

SPELLOW MILL.

This mill, also slightly beyond Liverpool boundaries, existed till the early part of the last century. It stood near the mere at Walton, in the neighbourhood of Spellow House. In 1803, on this latter being offered for sale, it was termed "Drysdale's Mill," and in 1813 it occurred again as "Drysdale's Mill on Walton Breck." In 1810 its lonely neighbourhood earned as unenviable a reputation as that of the Gallows Mills: "On Saturday evening (October 29) a gentleman on horseback was stopped by two men at Walton, near Spellow Mill; they led his horse twenty yards up the narrow lane leading to the mill and robbed him of his watch and some silver": and on December 17, the same year, a gentleman's servant was forced into the same lane and robbed, "the robbers then shook hands with the man and warned him not to leave the lane until they had got away." The mill, which was of the antique peg type, was burnt down in 1828.

CHAPTER V.

JEDBURGH CORPORATION MILLS.*

V. JEDBURGH CORPORATION MILLS.

1. Purchase, 1670.

1. THE story of the experience of the corporation of Jedburgh in running soke mills is chiefly interesting as depicting, first, the voluntary creation of a soke, imposed by burgesses upon themselves; and, second, the enforcement of regulations, unsuccessful in the end, for preventing the import of ready-ground flour and meal into the town.

The mills to which this little record refers were originally those of Jedburgh Abbey. They comprised corn mills and a fulling mill situated near the monastic house, and also closely adjacent to the borough. The corn mills were severally known as the Abbey Mill, the Town Mill, and the East Mill; the last named only being used for grinding wheat, and the others for reducing "hummel grain," or humble grain, *i.e.* oats, etc. The entire set of mills were "commonly called the Jedburgh Mills"; either, as some said, because they were in the suburbs of Jedburgh, or, as others contended, because the borough was astricted to them—an apparently trifling affair, but, in matter of fact, one from which some momentous issues ultimately sprung. After the Dissolution monastery and mills passed together to the Crown, and in March 1606 were granted by James to Alexander, Lord Hume; the charter being confirmed by an Act of Parliament in 1621. From

* The facts upon which this sketch has been founded appear in a printed report of an action-at-law in 1843, kindly lent by J. H. MacAdam, Esq., F.S.A., Scotland. Mr. McAdam is a studious and enthusiastic collector of books relating to the archæology of bread baking, and possesses a library of these quaint works—English and Continental—that, probably, is unique.

V. JEDBURGH CORPORATION MILLS.

1. Purchase, 1670.

Lord Hume the mills passed to Lord Lothian, from whom, in 1670, they were purchased by the Provost and Town Council of Jedburgh on behalf of the town.

The borough was deeply involved in debt, and the purchase of the mills constituted one of the methods of getting rid of the incubus which commended themselves to the authorities. The mills, therefore, were then a valuable property, rendering a definite and, as the council considered, a sure revenue, the profits from which ought to be devoted to the discharge of the local debt. It may be stated at once that, though no doubt burgesses ground at the mills, these do not seem to have ever possessed the thirl or soke of the town. No evidence occurs that the monks had ever compelled burgesses to grind at them, and the abbot was certainly not the manorial lord of the town, as it was a royal borough. Moreover, it does not appear that the corporation originally thought any such milling right existed, but simply resolved to buy the mills and do the best they could with them.

The purchase is thus recorded: "In 1670 William, Earl of Lothian, and Lord Robert Kerr, of Newbattle, his eldest son, sold and disposed to the provost, magistrates, and town council of the burgh of Jedburgh, on behalf of the community [*inter alia*], the three corn mills of Jedburgh and all and sundry multures, sequels, and others, and parts and pertinents of the same respectively"; no mention whatever being made of any astricted thirl over the burgesses of the town. The transaction was confirmed by royal charter in 1672, in which the property was described as "all and whole the mills denominated of old the three corn mills of Jedburgh, commonly called the Abbey Mill, Town Mill, and East Mill; as also the Waulk [walk or fulling] Mill of Jedburgh; with all and sundry multures, sequels, mill lands, houses, biggings, yards, and haill

V.
JEDBURGH CORPORATION MILLS.

1. Purchase, 1670.

parts and privileges of the same respectively ; with the parsonage teinds whatsoever of the same mill lands and others ; which subjects, including the said multures, sequels, and parts and pertinents, have been possessed and enjoyed from time immemorial by them and their predecessors in office on behalf foresaid." It seems evident, therefore, that the millowners then held no rights over the town. But the purchasers immediately set about creating such a right.

The corporation were compelled to borrow money to cover their purchase. According to a deed recorded in the sheriff-books of Roxburghshire, the provost, bailies, etc., convened a town's meeting February 5, 1670, for the purpose of considering and condescending upon what effectual course might be taken "to satisfy Mr. John Watt of Rosehill of the soum of 19,000 merks money as the aggried pryes of the thrie corne mylnes and waulk mylne and uthers, disponed be the Earle of Lothiane and Robert Lord Ker of Newbattle, quhilke is assigned be them to the said Maister Johne as ane of their creditors." At this meeting, numbering over two hundred burgesses, it was resolved that "they and everie ane of them most freelie, willinglie, and unanimouslie condiscendit, concludit, aggried, and be their prets condescends, concluds, and aggries in manner, form, and effect as after follows ; to wit, that the haill common goods, particularlie the Customs, Weyhts, Halfe-furlott, and Laidle of the said boroughe, with the thrie Corn Mylnes, Walkmylne, hous's, yards, lands, and outhers belonginge thairto, and also the ground annuals belonging to the said broughe, with the Counsell House, loft and lande and Boothes thairof, shall be rouped and fermed to any person or persons wha will undertake payment of the foresaid soume of 26,000 marks or thairby."

V. JEDBURGH CORPORATION MILLS.

2. Creating a Soke, 1670.

2. In this projected fee-farme lease of the whole of the common effects of the town, the mills constituted a conspicuous item, and it was to give practical effect to their value that the council proceeded at the same meeting to bind the burgesses, with their consent, to grind at them, thus for the first time creating a soke or thirl over the town of Jedburgh.

For the better improving of the said Mylnes, the foirnamed haill inhabitants and everie ane of them has thirled, astricted, and heirby thirles and astricts them and everie of them their airs and successors to grind at the saids thrie corn mylnes or any uther mylnes qlk sall happen to be erected or built be the said broughe of Jedbrughe in any part of the water of Jed, betwixt the Towerburyfoot on the eist and the landes called the Virgine on the west, all and quhatsomever, particularlie wheat, beands, peis, ry, beir, aits, malt or any uther sort of grain qever quilk they sall make use of; and that yeirlie, continuallie, and perpetuallie in all tyme coming.

And that they sall not abstract the samen cornes nor any pairt thairof from the said mylnes, but sall grind and suffer the same to be ground thairat; and sall pay the multors, knafesips, and uther deuties underwritten thairfor, viz. :—

for ilk boll of clad shilling, thrie capfulls of multor; or for ilk six furlots, ane peck; and for the knafesipe of ilk boll of meill, twa capfulls.

Item. Ilk boll of wheat, twa capfulls of multor, and for the knafesipe thairof ane capfull of flour.

Item. Ilk boll of humble corn, peis, beinds, and ry, to rait all alyke to the boll of wheatt.

Item. Ilk half boll or thrie fulls pay ane capfull of multor; or six fulls of malt, twa capfuls; and six furlots or nyne fulls, thrie capfuls; and a laid, ane peck.

And that under the paine of payment of doubell moultor of whatever cornes they sall abstract; or such soumes as the Toun Counsell from tyme to tyme sall think fit to modifie for abstracting of ilk boll of victuall from the saids mylnes.

The tacksmen, fermorers of the saids Mylnes sall, upon thair owne expensse, carie and transport the malt and cornes of the haill inhabitants from thair houses, and uther pairts within and adjacent to the said broughe, to and fra the said mylns; and sall deullie and reddilie answeir everie inhabitant as they sall be requyred upon tymous advertisment; and sall grind everie one's seck, pock, or others according to thair several advertisments and incoming, without respect of persons.

The Toune Counsell and others foresaids sall not ferme or roupe the said mylnes, nor grant taks thairof to any person or persons without it be speciallie provydet be them, as it is hereby speciallie provydet be the haill inhabitants abouenamed and underscryband.

V. JEDBURGH CORPORATION MILLS.

2. Creating a Soke, 1670.

The entire community having thus agreed to be astricted to the mills, the corporation duly leased them out, and upon this footing matters stood for nearly two centuries. On March 11, 1730, the magistrates and council "resolved that they would expose their milnes to roup on the 26th inst, and appointed publeck intimation yrof to be made at the Mercat Crosse and Kerk Door, advertising inhabitants and Strangers to come and bid at ye said roup: And recommended to the magrats or any of the council as a committee, three being a quorum, to meet on the 14th inst. and concert proper rules and articles for rouping of the said milnes and settle the conditions to be performed, hinc inde, by the inhabitants or persons that shall farme the said milns." On March 25 they resolved that "it would be for the benefit and advantage of the entire inhabitants" that the mills should be set in tack for nine years; and ordained, among other regulations, that "no other dues be taken at the milns, but according to the regulation in the Act of Thirledge of February 5, 1670 (ane attested copie wherof to be given to the lacksmen in order to determine all disputs), any custom or practice in the miln to the contrair, notwithstanding."

In 1740 the first symptom of public discontent with the thirl seems to have arisen, and, it being said that a rumour was gaining currency to the effect that certain papers were in the hands of the town clerk, "whereby it is surmised the inhabitants are free of stent," the clerk was directed to produce the said papers, though upon production they proved to be nothing more than various of the foregoing contracts of the council for leasing the mills.

3. Import of Meal prohibited, 1754.

3. The crucial question, and that upon which the thirl was ultimately broken, arose in 1754 in the form of the opposition of the council and the millers to

V. JEDBURGH CORPORATION MILLS.

3. Import of Meal prohibited, 1754.

the importation of meal already ground, for which, of course, the purchasers refused to pay any grinding toll to the town mills. It will be noticed that on this matter the council, as mill-owners, were somewhat weak; their original deed of purchase in 1670 had contained no clause binding the town to grind at the mills, nor (as was frequent in such clauses) prohibiting the import of meal, etc., already ground; nor, again, in the agreement or contract of February 5, 1670, creating the thirl, was there any such prohibition. Yet when, in 1754, Isaac Wright, Ninian Black, and other private burgesses, not bakers, imported meal, the council entered actions against them. They pleaded that these burgesses "have several times within two years not only abstracted certain multures of corn from the mill, which they have carried to other mills and ground there, but also, with a palpable design and intention to elude the effect of the thirlage, do barter their oats and other grain for meal; or otherways do sell their corns and buy meal without the thirl, and bring it in and consume it in their families without paying multures or customs due for each boll of meal brought within the thirl; to the great prejudice of the revenue of the town and the serious loss of the tacksmen of the mills and customs."

The result of the action does not appear, but in 1759 the council formally resolved that, "Whereas it was observed by the provost that there has happened several debates betwixt the tacksmen of the milns and the inhabitants with respect to importing meal into the town which had been grinded at other milns, and only paid eightpence at the boll [at those mills], it would be expedient, in order to prevent which, that some persons of the council should be named to enquire into the several acts of council anent the thirlage and importing meal into the town, and to prepare an

V.
JEDBURGH CORPORATION MILLS.

3. Import of Meal prohibited 1754.

overture for explaining and amending such acts, in order to promote peace and establish a rule whereby justice may be done both to the tacksmen and the inhabitants." Doubtless the committee found that the council had no prescriptive customary right to prevent the import of meal, but no result of their labours is recorded.

In February 1761 the council, still in a quandary, vaguely resolved that, "taking into consideration that the acts and regulations for levying dues at the milns and gathering the customs and other branches of the town's revenue are fallen into confusion by not being duly transcribed in the dean of guild book," a committee should "make up a statement of the acts and regulations, old and new, with power to make what improvements they shall think most conducive for the interest of the town and benefit of the inhabitants." In March, the committee having examined the said acts "and made some alterations therein," the whole were adopted and entered in the guild book, and duly signed by the provost on every page. The upshot was that "the haill inhabitants within the said burgh, being astricted and thirled to the milns of Jedburgh, now comprehending the barley and walk milns, conformable to the contract of thirlage of February 5, 1670, shall in all time coming pay and perform the multures and services following"; the multures being scheduled as already given.

Three years later the difficult problem at issue was attempted to be solved by a simple enactment of the council (March 1764) that, "Whoever of the inhabitants shall grind any corns, whether growing on the Jedburgh lands or elsewhere, at any other miln than the Jedburgh milns, shall be lyable in double multure whether the meal be grinded for their own private use or be sold by them within the town." But obviously,

V. JEDBURGH CORPORATION MILLS.

3. Import of Meal prohibited, 1754.

if it were not open to the inhabitants to grind grain elsewhere themselves, they were still left at liberty to purchase flour that other persons had ground outside the town; and after three years' experience of this state of affairs, the council made bold to suppress it by a distinct enactment, dated April 12, 1777:—

> It being represented to the council that the bakers of this burgh are beginning either to buy wheat in the country and get the same grinded at other mills, or to buy flour at other mills and immediately import the same into the burgh for the consumption thereof, to the great prejudice of the tacksman of the mills; and the council, being informed that no such practice ever yet obtained in this place since the constitution of the thirlage, and having observed that if some proper check was not immediately put to such proceedings, so unwarrantable and so destructive of the public revenues of the burgh, it might prove of very dangerous consequence: All which being taken into serious consideration by the council, they unanimously resolve that these or other devices of the like kind are unlawful violations of the rights of thirlage, to which the bakers and all the other inhabitants of this burgh are liable and obliged; and they do prohibit and discharge all and every baker within this burgh or suburbs thereof from consuming or making into bread for sale any flour or wheat meal of any kind but such as has been grinded and paid multures at the flour mill of this burgh, to which they and all the other inhabitants are astricted; with certification that should they do to the contrary they shall not only be liable for double multures for every boll of flour they shall so import, and proportionately in any such brought against them by the tacksman of the mills, but, further, shall be liable to the confiscation of the flour.

In 1799 the mill-toll in kind was converted into a money payment of 3s. 6d. per boll. In 1822 this was altered into 2s. 6d. per boll, when the price of wheat, by the monthly returns, was below 30s. per boll; 3s. when wheat was between 30s. and 40s.; and 3s. 6d. when wheat stood at 40s. and upwards. At the same time a number of the bakers compounded with the then tacksman of the mills; so that, on their engaging not to import any flour into the borough, but to grind all their grain in the town, they should pay so long as his tack lasted only 3s. per boll. The council agreed to this, and added the proviso that, "if the bakers should find

V. JEDBURGH CORPORATION MILLS.

it more for their advantage to buy flour than wheat, it is to be understood that they are to pay the 3s. for each bag of flour so imported into the burgh by them."

4. Soke abolished by Law, 1843.

4. It was not till 1840 that this and other regulations enacted by the council were seriously disputed. In that year action was taken against five bakers and meal dealers—John Madder, Andrew Hope, George Young, Robert Walker, and Alexander Balfour—who "not only abstracted and withheld their grain from the mills, but were in the practice of bringing into the burgh, both in manufactured and unmanufactured state, and also for their own private consumption, large quantities of grain, flour, and meal; and likewise, in defraud of the thirlage, were in the practice of purchasing grain of every description without the burgh, and also by sample in the Jedburgh markets, and getting the same ground at other mills; and thereafter carrying the same into the burgh without paying multures to the tacksman of the Jedburgh mills." The case was tried before the Lord Justice Clerk in January 1843; when the defenders pleaded (1) that the acts of the council did not constitute a thirlage, and there was no evidence that at the date of the purchase of the mills in 1670 the town was thirled to the mills; (2) that a commutation of the toll to money payment in 1799 had not endured sufficiently long to establish it as a custom; this latter being a side issue, having ultimately no bearing on the award of the court, and being of no present interest. In support of the case of the Corporation against the import of flour and meal, various evidence was taken, some brief extracts from which are of interest:—

Richard Christie, baker.—Did you hear of such a thing as the bakers bringing in flour without the miller knowing it? Oh yes, I

V. JEDBURGH CORPORATION MILLS.

4. Soke abolished by Law, 1843.

heard of it.—Sometimes they tried to bring in a cargo of flour overnight? Yes.—Secretly? Yes.—Without notice to the tacksman? Yes.—Did they always escape? Not always.—The tacksman was as sharp as they were? Aye, was he.—The tacksman, Gray, was pretty sharp? Yes, sometimes he had people for the purpose of watching the avenues to the town.—What was the result? They were taken before the provost and had to pay the dry multure.—What was dry multure? As much as they paid for grinding at the mills.—They paid for the flour as though they had ground it at the mill, although it was not ground there? Yes.—That was all they made by trying to bring it in in secret? Yes.

John Currie, carter.—Were you at one time employed a good deal by the bakers of Jedburgh? Occasionally.—Where did you cart the flour from? Different parts out of the town.—Did you bring it in openly? I brought it in at all times.—Did you ever bring it in the night time? I have.—Why? I do not know: whenever they bade me go, I did as I was bid.—Was no reason given for bringing it in at night? Not to me.—Did you know the tacksmen of the mills were watching for you at night? Yes, I knew they were.—Why were they watching you? I had nothing to do with that; they have attacked me on the road, but I cared nothing about it; they came to us and said they had catched us now.—And what did you say? Nothing but "come and get it"; I had nothing to do with it: it was not my stuff.—Did the millers get it from the cart? No, I went straight home with it to the man that it belonged to; the tacksman of the mill came up and had some words about it; I had nothing to do with him, and delivered my goods.

John Young, baker.—Was the charge for grinding at the country mills less than at the town mills? It was 2*s.* a boll less, I think.

The Lord Justice Clerk definitely directed the jury that the agreement made on February 1670 at the meeting of the council and burgesses was insufficient in law to impose for the first time, for all time coming, on the royal burgh and its inhabitants the restrictions and servitude of thirlage to mills acquired by the magistrates by purchase. A verdict by the jury in favour of the defenders disposed of the whole case; decree being ultimately pronounced against the corporation, who were cast in entire costs.

The thirl of Jedburgh mills was no sooner thus broken, and their value to the town lost, than four of the late defendants published a report of the trial sarcastically dedicated to the council and magistrates

V.
JEDBURGH CORPORATION MILLS.

4. Soke abolished by Law, 1843.

"as a lasting monument to their wisdom and prudence." In their preface they state that in 1839 they had offered to the corporation to take the opinion of counsel as to whether any thirlage really existed or not; that upon this being refused, they offered to rent the mills from year to year, or upon a lease of £350 a year; and on this being refused, to grind "all the wheat they required in the way of trade" at the mills at the usual rate of 3s., and to pay 2s. 3d. for "every boll of flour imported by them into the burgh." Such an arrangement, needless to say, would have resulted in the bakers finding they needed flour and not wheat in the way of trade, and this proposal also was rejected by the council. The late defendants conclude their printed report with the following: "The defenders do not mean to say that the verdict in their case ruined the burgh: all they wish to say is that it merely brought on critical days; the burgh has been in a bankrupt state for a hundred years at least; the conduct of its rulers has been the same in every age, and it is to be feared will continue so till the present system give place to a free and healthy competition." This, then, was the close of nearly two centuries' labour of the Jedburgh council to prevent the importation of flour into the milling preserve they had endeavoured to create.

GENERAL INDEX.

FINIS.